Virtual Reality, Empathy and Ethics

Matthew Cotton

Virtual Reality, Empathy and Ethics

palgrave
macmillan

Matthew Cotton
School of Social Sciences, Humanities & Law
Teesside University
Middlesbrough, UK

ISBN 978-3-030-72906-6 ISBN 978-3-030-72907-3 (eBook)
https://doi.org/10.1007/978-3-030-72907-3

This Palgrave Macmillan imprint is published by the registered company Springer Nature Switzerland AG.
The registered company address is: Gewerbestrasse 11, 6330 Cham, Switzerland

Acknowledgements

Thanks go to my colleagues at the University of York and Teesside University for providing a supportive atmosphere for the research to take place; to the editors and reviewers at Palgrave/Springer-Nature for their insightful comments and advice; and to my family—Helen, Faye, Bethany, Michael and Debbie Cotton for helping me to complete this book during a pandemic.

Contents

Abbreviations

ABJ	A Breathtaking Journey
ATE	Anticipatory technology ethics
ATL	Across the Line
AR	Augmented Reality
BBC	British Broadcasting Corporation
CA	Cambridge Analytica™
COS	Clouds Over Sidra
GIS	Geographic information system
GPS	Global positioning system
HILA	Hunger in Los Angeles
IDC	International Data Corporation
MMORPG	Massive Multiplayer Online Roleplaying Game
NGO	Non-governmental organisation
NLP	Natural language processing
OTA	Office of Technology Assessment
PDA	Personal digital assistant
POST	Parliamentary Office of Science and Technology Assessment
PTSD	Post-traumatic stress disorder
RPG	Roleplaying Game
RRI	Responsible research and innovation
SCOT	Social control of technology
SDG	Sustainable Development Goal
SDGAC	Sustainable Development Goal Action Campaign
SECT	Socially and ethically contentious technology
TA	Technology Assessment
TMTBA	The Machine to Be Another
UNMC	United Nations Millennium Campaign

UNVR	United Nations Virtual Reality Series
VIEW	Virtual Interface Environment Workstation
VR	Virtual Reality
WEF	World Economic Forum
WWW	World Wide Web

List of Figures

CHAPTER 1

Virtual Reality

Abstract In this introductory chapter, virtual reality (VR) is discussed as both concept and modern technological artefact. A brief history of virtual reality is documented—detailing innovations in stereoscopic vision, auditory and haptic feedback, motion detection, and computer-generated models. The factors leading to a growing contemporary market for low-cost domestic VR technologies are discussed. The range of applications for VR and augmented reality (AR) across education, architecture, training, cultural heritage, public engagement, and medical therapy are assessed. The chapter concludes by assessing the potential benefits and disadvantages of VR technology-application to different fields.

Keywords Virtual reality • Augmented reality • Technology markets

Introduction

The objective of this book is to assess the ethical dimensions of virtual reality (VR) in society, and then to proffer the technology as an ethical tool—a means to assist in ethics training and decision-making by structuring and simplifying user engagement with moral choices. The analysis presented draws together insight from applied philosophy, science and technology studies, and journalistic practice in the construction of such a tool. The structure of the book is as follows. Chapter 1 provides a brief

M. Cotton, *Virtual Reality, Empathy and Ethics*,
https://doi.org/10.1007/978-3-030-72907-3_1

history and overview of VR. It explores the successive technological advances deployed to stimulate a sense of immersion and embodiment in users. It then reviews the application of contemporary VR technologies in a range of fields, including medicine, architecture and military training. Chapter 2 explores the broader application and social impact of VR technologies, including what I term 'covert virtual objects' such as chatbots and 'deep fakes' that present a growing threat to social cohesion and information exchange across social media. The ethical dimensions of virtual communities are discussed, as well as the potential benefits that can be derived from stimulating prosocial behaviour through engagement in virtual environments. Chapter 3 discusses the idea of ethical tools—reviewing a range of decision heuristics and judgement aids to assist moral reasoning, providing insights from applied ethics research across a range of disciplines. Chapter 4 examines the role of empathy-arousal in virtual environments, and the benefits and drawbacks of taking an empathic approach to ethical decision-making. Chapter 5 then presents a range of practical cases of VR technology use in empathy arousal, including the United Nations' VR program, drawing insight from these examples to the development of a new ethical tool. Chapter 6 concludes by bridging VR-based empathy arousal projects and ethical decision-making tools into a coherent framework and provides an example of how such an approach could be used in medical ethics training. Throughout the book, reference is made to pragmatist philosopher John Dewey's work on 'dramatic rehearsal': that by combining empathy-arousal, moral imagination and structured role-play within a virtual environment, the technology platform provides multiple benefits in training and decision-support for a range of applications in practical ethics.

VR is commonly understood as a computer-oriented artefact, or technologically mediated space. VR is a place that one goes *to* and experiences by putting on a headset and interacting with filmed or computer-generated objects and characters within a 360° field of vision that tracks with physical movement through space. The technology becomes something of a portal to an artificial place, one that Baudrillard suggests (Ryan, 2001; Baudrillard, 2005) is an illusion—a surface-level effect produced by actual physical causes and interactions at a material level. Etymologically, virtual, from the Latin *virtus*, means *power*, *force*, *capacity*, *ability* or *fact*; yet curiously in many languages that derive words from that root, *virtual* means precisely the opposite. The virtual is something that is not quite a fact, something that cannot be touched, and exerts no direct force. It can therefore be understood through the analogy of the *optical illusion*—it is a condition of existence which is not real but that displays many of the properties and qualities of the real. The simplest of these surface-level illusions is one's own reflection in a mirror. The reflection already exists in one sense, whether or not one can actively see it. Reflections are a physical phenomenon—they are the change in direction of a wave front (in this case of a photon) at an interface between two different media, such that the wave front is returned to the medium from which it originated. The specular reflection of a flat mirror allows an individual to see their own reflection as if it were a copy of themselves on the other side of the glass, instantly mimicking their own behaviour. However, sociologically, the image is something that must be encountered; the reflection is a virtual object that is ascribed meaning through psychological and social processes of interpretation, not just the physical laws of optics. It is this process of encountering and interpretation that makes the object virtual, and through which an understanding of a VR is constructed. The virtuality of the reflection is not solely passive and observed. The virtual is thus "real but not actual, ideal but not abstract" (Gaffney, 2010; Moulard-Leonard, 2008): that is, the virtual world belongs to the domain of subjectivity, perception and immateriality, and yet it is something that one experiences as if it were real. It is something that is *almost there*, but not quite. It is this blending at the interface of real and non-real objects through our sensory perception and mental interpretation through which a virtual reality gains social meaning. Encountering the virtual object has real effects upon the psyche of the individual, upon self-perception and the imagination of the image. We can therefore understand that even simple virtual objects like a reflection have both material (physical) and psycho-social causes and effects that extend into the 'real' world beyond the virtual space.

Though *virtuality* sits within a complex philosophical debate, VR has a more prosaic public identity. Shared social representations of VR are influenced by rapid technological advancement and the commercialisation of head-mounted displays, controller mechanisms and software platforms. These technologies are *branded* as VR; and so the concept has come to refer to the size, shape and use of certain technologies. Yet definitions of what is and isn't VR remain imprecise. VR is often promoted in media discourse as a novel and exciting form of computer interaction—for some the term is used as a shorthand for *any* type of three-dimensional computer presentation (Wann & Mon-Williams, 1996); for others it is used to connote a collection of specific types of computer hardware—usually involving some kind of head-mounted display, an input device (such as a glove or other hand-held controller) and a linked auditory stimulus. These are collectively spatial technologies—it is the physical movement of the control mechanisms through both real and imagined space that creates the virtual environment. When thinking of VR as a technology, it is the combination of hardware, software and movement that produces a specific 'reality' into which a user is immersed. Technologically mediated VR therefore works through a process of simultaneous stimulation and sensory deprivation in order to produce an artificially immersive sensory experience. The experience is only *virtual* if it sufficiently fools the brain into a sensation of being somewhere else. The reality is created in the mind of the user. Virtual realities can therefore be either solely individual, or they can be collectively shared with other users linked through networked computer systems. There can therefore be collective virtual realities—raising important questions concerning the social and ethical impacts of virtual reality as a *socio-techical* system. Reality 'inside' the experience is very different from 'outside' of it, and it is the interface between these two realms that produces social and ethical dilemmas discussed in Chap. 2.

A Brief History of Virtual Reality Technology Development

Technologies to enhance or alter individuals' experience of simulated realities are not altogether new. One of the earliest examples of a technologically facilitated virtual object was Sir Charles Wheatstone's invention of the 'stereoscope' in 1838. A stereoscope has two windows, one for each eye, each projecting a different picture. In one is a left-eye depiction of the scene and in the other, a right-eye depiction. The stereoscope technology has a lens for each eye that makes the image on each side appear both larger and more distant (and also shifts its apparent horizontal position). The effect is such that for a person with regular binocular depth perception, the edges of the two images seemingly blend together to form a single 'stereo window'—the sensation of a single three-dimensional image. This works by tricking the brain's interpretation of the optical sensory information that it receives, much in the way of an optical illusion. When the two images simulating left-eye and right-eye views of the same object are presented so that each eye sees only the image designed for it, seemingly in the same location, the brain will fuse the two and accept them as a view of one solid object (Brewster, 1856; Wade & Ono, 1985; Crone, 1992). Stereoscopes continue to this day, with the View-Master patented by William Gruber in 1939 still in production.

The growing popularity of the cinema in the 1950s led to a number of technological innovations (some more successful than others) to lure customers into theatres. Three-dimensional stereoscopic movies were a significant advance, based upon similar technology to Wheatstone's picture viewer. The illusion of three-dimensional moving pictures was derived from the same technique of stereoscopic photography, whereby a regular motion picture camera system is used to record the images as seen from two perspectives, then either special projection hardware or eyewear (commonly with two differently coloured lenses) is used to limit the visibility of each image to the viewer's left or right eye only, and as before, the brain is tricked into perceiving the images as containing virtual solid objects within a depth of field.

Changes to optical technologies to enhance the sense of immersion through perceived object depth were augmented with technologies to enhance other sensory components of the viewing experience. In early 1931, British audio engineer Alan Blumlein was critical of the audio reproduction systems of the early 'talky' films, which relied upon a single set of speakers for the projection of dialogue into the theatre. This led to the disconcerting effect of the actor being on one side of the screen whilst the actor's voice appeared to come from the other. Blumlein's modern *stereophonic* sound was invented as a means to reproduce sound using two or more independent audio channels and a symmetrical configuration of loudspeakers. This created a more natural impression of sound heard from various directions, as the structure of the audio environment more closely mirrors that of natural hearing (Torick, 1998; Paul, 2009). It also allowed a greater sense of immersion in the audio environment of the film, by allowing sound to seemingly move across the audio field as actors moved across the screen. By synching audio and visual environments the sensory experience is deeply enhanced, and innovation in multichannel sound to create greater depth within audio environments continues to this day.

With advances in visual and audio immersion, other technologies began to emerge to enhance the sensory experience further. Smell-O-Vision, developed by Hans Laube, was a technical system that released different odours in the movie theatre during the projection of a film. The idea is that, like a film's score enhancing the auditory sensory experience of the drama, if the viewer can smell what is happening in the movie this will further enhance the sense of immersion, heightening suspense and the dramatic effect of the images on screen. The system was used only once in practice, in the film Scent of Mystery in 1960. Smell-O-Vision employed 30 different odours, releasing them directly into the theatre, triggered in time to the film's soundtrack. It was not a success with moviegoers so was quickly abandoned (Cheok et al., 2013; McGee, 2001). However, the notion that stimulating multiple senses simultaneously as a means to enhance the effect of immersion, remained appealing to technology developers, with evidence that smell training through learning games holds promise as a means of improving cognitive function (Olofsson et al., 2017). Further research in this area is ongoing.

The *Sensorama* developed by cinematographer Morton Heilig in 1956 was an attempt to bring together multiple sensory stimuli within a small-scale theatre cabinet in order to produce an immersive moviegoing experience (Heilig, 1962). Heilig wanted people to feel like they were in the movie they were watching, rather than simply viewers of the experience. The Sensorama cabinet aimed to stimulate all the senses (not just sight and sound). It featured a stereoscopic display and stereo speakers, but also smell generators and fans to simulate wind, and a vibrating chair to simulate haptic feedback and movement (Dinh et al., 1999). A notable example of one of Heilig's films was a simulated city environment, in which the user rode through on a motorcycle. This early VR system used sensory stimulation to allow the user to experience the road, hear the engine, feel the vibration, and smell the motor's exhaust within a simulated virtual world. Heilig later patented a head-mounted display device, called the Telesphere Mask, in 1960, though notably it differed from the modern VR headset, in that the film medium was not interactive, and there was no motion tracking involved. The experience of the Sensorama was more than simply a passive watching of a movie—it was an active participatory experience within a prototypical virtual environment, and the development of the technology became foundational to later iterations of head-mounted VR displays.

In 1965 Ivan Sutherland wrote a paper outlining a concept note for the design of The Ultimate Display (Sutherland, 1965/2002)—one which could simulate a virtual environment with such accuracy to the point where the user could not tell the difference from the physical reality of their everyday lives. Sutherland (ibid.) describes the Ultimate Display VR technology as such:

> *The ultimate display would, of course, be a room within which the computer can control the existence of matter. A chair displayed in such a room would be good enough to sit in. Handcuffs displayed in such a room would be confining, and a bullet displayed in such a room would be fatal. With appropriate programming, such a display could literally be the Wonderland into which Alice walked.*

Though seemingly fantastical in Sutherland's time, the paper would become a core blueprint for the concepts that underpin VR today. Sutherland, along with his student Bob Sproull developed an early head-mounted display, termed the Sword of Damocles (the technology was too heavy to be worn on the head, and so was attached to the ceiling, thus giving it its name). The advancement was significant however, as the Sword of Damocles projected computer-generated images, rather than stereoscopic film. These images were simplistic wireframe rooms and objects, though this was clearly an emergent VR technology that provided an important precursor to what we understand as a modern 3D gaming environment.

Later advances in optic and haptic feedback devices in the 1970s and 1980s allowed a greater sense of movement around virtual spaces (such as NASA Ames Research Center's Virtual Interface Environment Workstation [VIEW]) system designed in the mid-1980s that combined a head-mounted, wide-angle, stereoscopic display system. This was controlled by a combination of operator position through motion tracking, voice and gesture using gloves to enable the haptic interaction (Franklin Institute, 2019). It was developed for use as a multipurpose interface environment. VIEW was multisensory and importantly involved user exploration of a 360° synthesised or remotely sensed environment. VIEW's primary applications were conceived as for tele-robotics, manufacturing, the management of large-scale integrated information systems, and human factors research (Fisher et al., 1988). The VIEW design format became synonymous with public perceptions of later VR technologies in commercial design and popular culture—and elements of this model are familiar in twenty-first century VR technologies.

Modern Virtual Reality Technologies

Conceptually, VR as a computer-mediated technology owes a debt to the work of Myron Krueger. Krueger wrote a thesis in 1974 which discussed the concept of artificial reality as a digital substitute to the real world. Krueger defined a number of core elements of such a reality, namely that it is *wholly computer generated*, is *interactive*, and that *immersibility is a fundamental characteristic* (Krueger & Wilson, 1985; Krueger, 1993). From the technological progress of VR up to the late 1980s, VR became synonymous with 3D computer technology designed to generate a sensory experience that simulates specific environments or 'worlds' for users. VR now commonly employs bespoke headsets or multi-projected environments (either alone or in combination with physical elements), to produce realistic images, sounds and often other sensations (including haptic, auditory or again there is a growing interest in olfactory feedback). The use of head-mounted displays reduces the cost of providing visual stimulus when compared to the augmentation of physical environments (large-scale display screens). Together, these stimuli produce a sensory environment which, when sufficiently immersive, simulates a user's presence in a different environment; breaking the connection between where our senses tell us we are and where we are actually located, and whom we are with (Sanchez-Vives & Slater, 2005). The immersive nature of VR technology is achieved by allowing a person to look and/or move within the sensory environment and interact with it. This interaction requires motion tracking technology—the technologies must sense where the headset/gloves/controllers and so on are in physical space and then map a corresponding location within the virtual environment. Interaction within the virtual environment could be with computer-generated objects and characters, or with other human agents within multi-user environments. Sometimes as with other forms of computer-generated world-building, virtual realities are limited only by the imagination of the designer and the user, VR technology therefore enhances the *experience* and impact to the user through the combination of imaginative world-building and bespoke immersive and multisensory interfaces.

VR technologies are now affordable as mainstream consumer technologies, and their types and applications for both domestic and commercial use are increasing. VR headsets for consumer purchase come in a number of different forms and formats. Autonomous VR headsets contain visual processing and computing hardware to create the simulated reality as a stand-alone piece of equipment. 'Wired' headsets connect to an external processing unit (usually a computer or games console) and only reproduce (rather than process) the visual/auditory data. As computing power increases within system-on-a-chip designs, the visual processing power of stand-alone machines is increasing, making wireless operation cheap and high quality. One of the largest growth areas is screenless viewers that house a high-spec mobile phone. At the time of writing, phones have screens large enough (roughly six-inch diagonal) and of sufficiently high definition (1080p equivalent or higher) to provide good quality binocular visual. Moreover, phones are equipped with accelerometers which provide Global Positioning System (GPS) information which can create corresponding movement of the visual as the device moves with head-tracking. Homemade VR kits such as Google Cardboard Viewer™—allow individuals to produce a low-cost VR experience using their phone. Given the ubiquity of mobiles amongst target consumer groups, these technologies provide mass accessibility of VR. As such, the technology is seeing rapid uptake in consumer markets. The International Data Corporation (IDC) predicts an annual increased sales rate of augmented and VR technologies of 66.3%, culminating in predicted global sales reaching 68.6 million units by 2023 (of which 36.7 million are VR headsets). Autonomous VR headsets (not requiring a connection to a PC or console) represent 59% of predicted sales, followed by wired headsets VR (connected to either a PC or console) with 37.4% of market share, whereas cordless headsets (screenless viewers) represent 37.4% of predicted market share (IDC, 2018). The steady growth of the VR headset market is shown in Fig. 1.1).

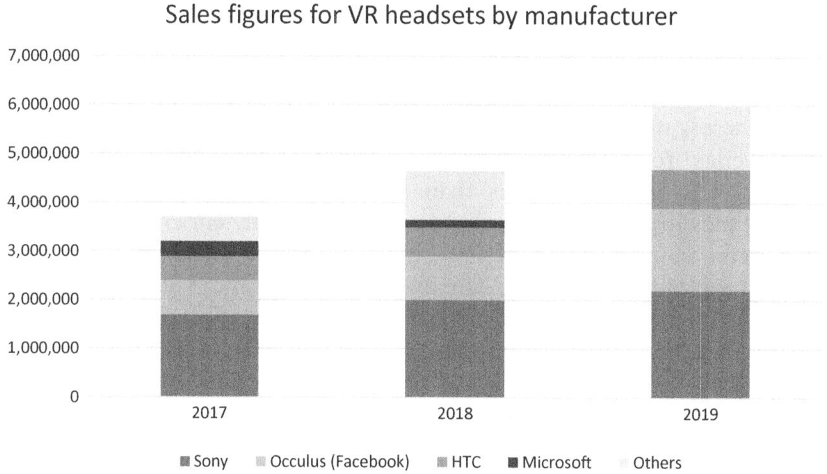

Fig. 1.1 Sales of virtual reality headsets by manufacturer (Source: Liu (2019))

The growth in consumer uptake of VR has a number of causes, functions and effects. For some, VR is a novelty—low-cost consumer technologies such as the Google Cardboard Viewer™, allow users to experiment with VR applications and videos for a range of user-recreational purposes—primarily gaming, but also films, 3D photography or painting (using, e.g. Google Tilt Brush™) documentaries, immersive sports entertainment (such as Intel® True VR which sets up cameras within sports grounds, and the feed is then relayed to headset displays to give a sense of being in the stadium itself). VR is even used in pornography. Though VR is a growth market, companies such as Google have a strong interest not in the traditional VR as a self-contained simulation, but in *augmented* reality devices and applications. Augmented reality is really a *blended* or *mixed* reality, created through technologically enhanced visualisation of existing physical environments. AR fits within what is sometimes described as a reality-virtuality continuum (Milgram et al., 1995). At one end of the continuum are computer-generated window-on-a-world systems accessed through computer monitors, tablet computers, personal digital assistants (PDAs) or smartphones, in which the screen displays a 3D and navigable world, though there is very little immersion beyond the direct visual and ambient auditory stimulus (stage 1 on Fig. 1.2). AR sits somewhere in the middle of this continuum. Like VR it is computer-generated and interactive, though unlike VR where objects and environments are completely computer-generated, in AR many real-world objects remain intact. Computer-generated perceptual information (usually visual but sometimes across multiple sensory modalities, including haptic and olfactory) are overlaid to real images (type 2 on Fig. 1.2). The aim of augmented reality is to enhance rather than replace environmental information. Sometimes this can be used to create a perceptually enriched experience for users—whereby components of the digital world blend into a person's perception of the real world. The digital information can also be manipulated, so by adding computer-mediated vision by incorporating augmented reality cameras into smart phones or wearable technologies (such as the somewhat ill-fated Google Glass™) augmented reality software can use object recognition to extract information about the surrounding real world and then allow it to become interactive for the user. Augmented reality therefore has considerable use in engineering, repair, inventory management or other real-time control of physical space that requires rapid access to spatially oriented data. By using AR cameras to scan or view images or marker-less objects the real-time display of spatially mapped

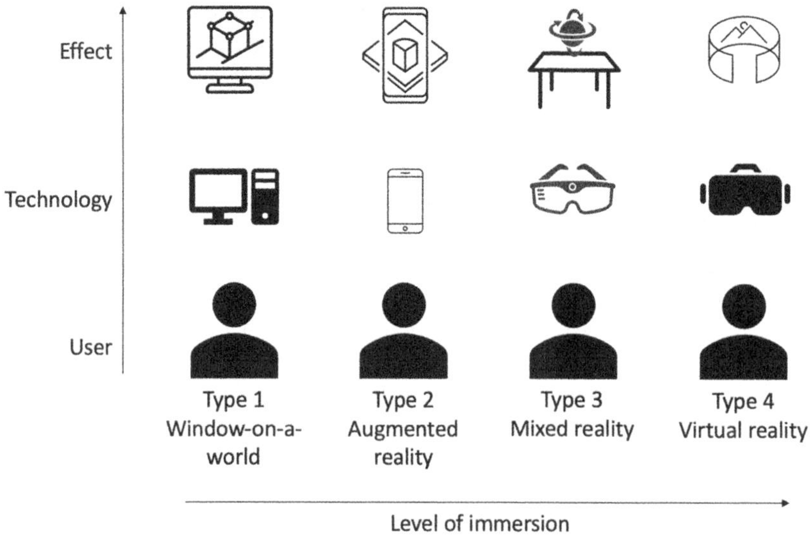

Fig. 1.2 A continuum of computer-mediated realities

information (using, e.g. geographical information systems (GIS)) the AR system can show the alignment of quantitative, spatial or physically imperceptible information to the user (such as showing electromagnetic wave information aligned with the physical space) (Mann, 2015).

Though there are numerous professional applications of augmented reality, the technology entered consumer mainstream consciousness with the arrival of the app Pokémon Go™ by Niantic. The smartphone game uses the inbuilt camera, accelerometer and global positioning system to allow users to explore their local environment using augmented map to find Pokémon™ (animal-like fantasy creatures that 'evolve' from one state to another by advancing within the game). Once they are found on the map, the app uses the smartphone camera to overlay 3D computer-generated image of the Pokémon onto video of the physical location behind it—this gives the illusion that the Pokémon is in front of the user in the environment in which it is found and then 'caught' by the user. The app developed by Niantic for Android™ and iOS™ devices was highly popular—with 500 million downloads worldwide by the end of 2016 (the year of its release). The popularity of the game spurred Apple IOS™ and Google Android ™ to release augmented reality developer kits, spurring a range of different augmented reality apps—for example, Just A Line ™, which allows users to draw on still images and videos.

New developments such as Sandbox VR™ are more sophisticated gaming systems (dubbed 'hyperreality' or sometimes 'mixed reality'—type 3 on Fig. 1.2) that combine headset-wearing users with motion tracking sensor suits and physical objects and controllers to create shared VR experiences within a fixed physical environment. This blending of virtual and real features, physical movement through space (thus involving proprioception as well as visual/auditory perception) creates a holistic sense of a different reality as virtual objects are manipulated in concert with real ones. Thus, mixed reality, like augmented reality shares common features of a non-computer-generated reality. It is only when the sensory input of real objects is excluded entirely from the user (by blocking out as much sensory input from the real world as possible), that VR at the far end of the continuum (type 4 in Fig. 1.2) is achieved. In simple terms, therefore, the greater the level of stimulation and integration of sensory inputs, the greater the experience of immersion.

Applications of Virtual Reality Technologies

The immersive nature of virtual and mixed reality technologies enhances the experience of first-person games such as shooting, driving, exploration, role play or simulators. Though gaming is driving the growth of headset sales, there are a number of other applications that are significant. As the scope of VR research and development has expanded beyond experimental/academic applications towards commercial realisation of technology projects, multiple industry sectors have made significant investments in R&D and manufacturing capacity development—not just the information technology sector, but also biomedical engineering, structural design and the training aids technology sector (Lele, 2013). In turn, the application of novel VR technology platforms is of growing interest to the fields of architecture and urban design, clinical settings and military training. Below are some examples of the technology applied to different fields:

Architecture and planning—VR like computer-assisted design (CAD) has its roots in visual communication science, and so has many applications to both core design and design communication fields. It has proven highly useful for communicating visual design disciplines—specifically in architecture, urban design and master planning. VR allows one to move around the virtual design of a hypothetical physical space. The VR programme can provide more than just a visual representation of the space—when programmed with parameters for physical features, architects and designers can better understand factors as varied as sight lines, thermal comfort from solar heat gain, material stress, or transition of materials. The use of the technology to physically model changes before construction has obvious benefits in terms of reducing cost and delays due to faults or changes at the construction stage. Unlike virtual environments used in health sciences or engineering, simulations using VR theatres or labs are commonly used to display physically *inaccessible* realities (i.e. planned and designed realities, not yet existent or with non-existent components) (Portman et al., 2015). There is a fundamentally *imaginative* element to such VR projects—they either stimulate the imagination of physical spatial reality, or substitute imagination with realistic simulation. VR can be employed to somehow make up for the difficulties of spatial imagination that clients, funders or other stakeholders have in visualising new or future projects.

Medical training—VR has long been heralded as leading to a paradigm shift in medical training and clinical practice—from VR surgical simulators, telepresence surgery, complex medical database visualisation, and rehabilitation (Seymour et al., 2002; Willaert et al., 2012). VR can be combined with mechanical devices such that teaching and evaluation of surgical skills can be done through proxy representations of the human body (Reznick & MacRae, 2006). There is growing enthusiasm in university hospitals for VR- and AR-based learning programmes. There are many different applications. Some are based upon microscale biology and anatomy education—such as programmes which allow users to explore neuroanatomy, the function of specific organs or the impact of biochemical changes in the body. Others allow training in specific surgical tools and techniques with real-time AR feedback. At the macro-scale, VR simulators can model real-life emergency medicine situations with not only a virtual patient, but also other medical practitioners as virtual characters. Though there is strong enthusiasm for the technology, medical educators are wary of placing too much emphasis upon virtual skills, given how little data there is on the efficacy of virtual training to prepare physicians and other medical practitioners for real-world scenarios.

Therapeutic applications—Though the value of VR as a training tool for medical practitioners needs further evidence, VR has been more thoroughly researched in a therapeutic context. VR has been applied and tested across a range of physiological and mental health issues. It has myriad applications to patient care, rehabilitation, comfort and rejuvenation. Often these concern pain management, for example using distracting immersive VR to reduce pain in post-operative rehabilitation, surgery or burn care (Hoffman et al., 2011; Hoffman et al., 2000), or rehabilitation or cognitive function preservation in patients with dementia (Garcia-Betances et al., 2015). VR is also potentially useful in the clinical treatment of mental health problems. As Freeman et al. (2017) have shown, mental health problems are effectively inseparable from the environmental conditions that the sufferer experiences. VR allows individuals repeated exposure to problematic situations and can facilitate them being taught, via evidence-based psychological treatments, how to overcome the difficulties that they experience in these situations. This is essentially described as 'exposure therapy'. Exposure therapy is effective in treating post-traumatic stress disorder faced by military veterans (Reger et al., 2011; Rizzo et al., 2005; Botella et al., 2015), the treatment of panic and anxiety disorders, and phobias (Powers & Emmelkamp, 2008). The sensory stimulation of the environmental agent that simulates mental distress can be conducted within a safe clinical environment. This has proven an effective treatment. Notably, Rothbaum et al. (2000) compared virtual and real exposure therapy to fear of flying and found that experience of a virtual plane is just as effective as experience of a real one in reducing fear of flying; and Emmelkamp et al. (2002) found it similarly effective for treating acrophobia (extreme fear of heights).

Military applications—Aside from the aforementioned recuperative benefits of VR to treat post-traumatic stress disorder (PTSD), it is also of significant interest to investors given its capacity to reduce military personnel exposure to hazards and to increase stealth (Herrero & De Antonio, 2005). R&D into realistic warfare simulators has been the primary focus of VR application to military training (Lele, 2013). Likewise, other dangerous professions such as mining (Orr et al., 2009) or fire safety (Cha et al., 2012; Kinateder et al., 2014) have made use of VR in training. Other than risk management from physical hazards, VR is also useful for providing low-cost training in skill development such as through man–machine digital interfaces to improve real-world firing range training (Bhagat et al., 2016), or else to improve overall performance, for example to train soldiers to deal with acute stress response under battlefield conditions (Rizzo et al., 2012).

Cultural heritage, museums and science education—Pujol (2004) notes that the social role of the museum has been changing in the face of technological advancement. Museums are exploring new interactive means of communication in order to better serve a young technologically savvy audience, yet the aim of museums as curatorial spaces for artefacts and cultural heritage is still a largely traditional one. Museums have traditionally been slow to adopt new technologies, yet they play a central role in making cultural heritage accessible to a mass audience. As such, in recent years curators have explored new means of technologically mediated cultural communication (Carrozzino & Bergamasco, 2010): a concept sometimes referred to as *edutainment* (Lepouras & Vassilakis, 2004). VR and mixed/immersive reality technologies (type 3 on the continuum in Fig. 1.2) can assist in this edutainment process by building a virtual museum structure, presenting 3D reconstructions of archaeological sites/artefacts or to reproduce virtual living ancestors from digital reconstructions of human remains. One notable example is the BBC-sponsored Civilisations AR app for IOS and Android, which ties into the Civilisations documentary series. The app uses AR to simulate the handling of 30 different artefacts—either viewed as lifelike 3D renderings through a smartphone (as viewfinder), or through other interactive features such as 'x-ray views'. Other examples are more hands-on. For example, in physical science education, an augmented reality sandbox allows users to create topographic models by shaping real 'kinetic sand'. The sandbox exhibit is augmented in real time by the projection of a colour elevation map and contour lines which exactly match the sand topography, using a closed loop of a 3D camera, simulation and visualisation software, and a data projector. When an object (such as a hand) is sensed at a particular height above the sand surface, virtual rain appears as a blue visualisation on the surface, and a flow simulation moves the water across the landscape (Reed et al., 2014).

Virtual, augmented and mixed reality applications have the benefit of allowing individuals to access the sensory detail of artefacts, objects or even landscapes which might otherwise be off-limits or behind glass due to their delicate nature, or else difficult to imagine, such as eroded archaeological sites, watersheds or decomposed human remains. VR has the potential to bridge an 'imagination gap' between the present and the past—by using mathematically driven reconstructions of damaged or decomposed items and then presenting them in immersive virtual visualisation; this has the potential to make the past come alive for lay museum users that are unfamiliar with the historical context in which the artefacts or sites were first produced. It is this blending of a scientific model of artefact reconstruction with the creativity of visualisation, that VR has particular power as a communicative tool.

Conclusions

The concept of VR has become synonymous with a type of computer-generated world-building that allows users to enter situations that they would not (or should not) otherwise experience. Since the early twentieth century inventors have been motivated to improve user experiences of media by ever increasing levels of immersion, first in cinema and then in computer simulation. From stereoscopic imagery, stereophonic sound and haptic feedback, technological programmes to create artificial reality have nonetheless remained specialist, niche products. In the early 1990s a flurry of interest in VR technologies lead to ever-growing public expectations about the state of the technology, its influence upon social life and upon the ethics of social interaction in virtual spaces (Kennedy et al., 2010). These issues are discussed in the following chapter. However, public expectations of the power of virtual worlds in the 1990s were let down by the limitations both of the technology and of human physiology. Krueger's (1993) promise of an immersive artificial reality remained unrealised for three reasons. Firstly, because computing power and environmental design was insufficient to build a rich and expansive virtual experience. VR remained a poor imitation of reality through the 1990s. However, as computers develop ever greater 3D graphics processing power, as well as high-definition optical displays, accelerometers and tracking sensors, the promise of immersive VR is growing closer to Krueger's artificial reality-ideal. Secondly, the costs of VR technologies have, until recently, remained prohibitively high. The independent Oculus™ platform for low-cost consumer VR headsets spurred a raft of activity in the technology sector, with other major players stepping in (including Facebook, which ultimately bought the Oculus platform). Low-cost adaptations to modern smartphones increased the potential for VR software developers to produce content for low-cost consumer platforms. Thirdly, the physiological problems for many users of VR remain off-putting. VR headsets provide sensory input to auditory and optic nerves, but often do not provide sensory feedback to physiological systems involved in proprioception (those involved in perception or awareness of the position and movement of the body, such as the nerves that sense pressure in the feet, and the vestibular system or 'inner ear' which is involved in balance). When the brain senses that the body is moving through space (from the optical and auditory sensory input), but gets no feedback from the stationary body sitting in a chair or standing still, then this can cause extreme motion sickness, which often persists long after the VR experience has ended (Hettinger & Riccio, 1992; Kennedy et al., 2010). Together, the design, cost and physiological limitations of VR restricted the appeal of the technology until relatively recently.

There is also a public perception problem for VR developers. VR is often dismissed as a frivolous gaming activity; yet the myriad range of applications outlined above show the adaptability of computer-generated sensory content to achieve multiple psycho-social objectives—from therapy, to cultural communication, and to education and training. VR can simulate different aspects of the physical environmental and psychological state of users, thus influencing the understanding and experience of the real world. Though this has recreational applications in gaming, social media and advertising, there are potential learning benefits as well. Empirical research across a range of disciplines shows that learning in virtual environments can be beneficial to improving real-world performance.

The benefits of VR are fivefold. First, VR allows training in a safe environment. Many difficult and dangerous occupations from military, law enforcement, fire safety and search and rescue operatives, must place themselves in dangerous situations, or else, their actions might harm others (as in the case of medical practice). Virtual practice of complex skills in the relative safety of a virtual training environment is potentially beneficial when compared to physical environmental training. However, often the technology platforms that are adopted lack a sufficient evidence base to assure users that virtual skills training is as good as real-world skills training, and so we must remain cautious about blanket recommendations from manufacturers to use VR as a primary training tool.

Second, there are a number of cost and other savings (such as time and effort) that can be gained from manipulating simulated objects and environments when compared to physical objects. In architecture, the construction of scale-models has money, time and space costs that visualisation of computer-aided design models does not. When visualisation requires spatial manipulation (such as building design, archaeological site traversal, or engineering repair), virtual objects are cheaper to produce, easier to manipulate and also help to improve the overall material resource sustainability of projects and plans.

Third, VR generates the capacity to manipulate individual elements of models, objects or environments and thus change environmental conditions to stimulate different kinds of learning and psychological development. Teaching and training require user learning through trial, error, reflection and re-trial. Virtual objects, by virtue of their ability to be manipulated, vastly extend the range of options for teaching and learning. By combining teacher and learner feedback into virtual environment and object design, the capacity for learning is potentially improved.

Fourth, virtual environments, by virtue of their computer-generated nature, are reliable and reproduceable. From the training and educational point of view, reproducibility allows standardisation of training provision; it also allows users confidence in the replicability of results for repetitive tasks. For example, in surgical skills training, where similar tasks must be repeated with high levels of accuracy, the reliability of virtual environment training is a boon for educators and learners.

Fifth, VR has a range of communicative benefits. These benefits are primarily based upon the creativity and the immersive nature of the technology that together bridge the aforementioned 'imagination gap'. For example, lay users of VR technologies in museums will be able to imagine past environments, people and artefacts more easily than through pictures or written narrative descriptions, and they can be reproduced either within an immersive VR, or overlaid on a specific site or artefact using AR. Thus, where physical or sensory data is incomplete this can be supplemented within a virtual model. The creative dimensions of virtual world-building are returned to in Chap. 5, in the development of an imaginative VR-based tool for ethical reflection.

Finally, the development of VR technological platforms is grounded in an ethos of *technological optimism*—that is an artificial reality that is immersive has benefits both recreationally, in terms of the immersive enjoyment of fictional media such as films and games, and practically, in terms of the ability to simulate reality, to manipulate it, to present data spatially and to benefit learning, design and implementation in the real world. However, the promise of VR has remained of concern to politicians, social commentators and philosophers since the 1990s. Concerns have been raised about the changes to the offline world that might occur as we blend our understanding of reality with that of the virtual, and these are the issues discussed in Chap. 2.

References

Baudrillard, J. (2005). Violence of the virtual and integral reality. *International Journal of Baudrillard Studies, 2*(2), 1–16.

Bhagat, K. K., Liou, W.-K., & Chang, C.-Y. (2016). A cost-effective interactive 3D virtual reality system applied to military live firing training. *Virtual Reality, 20*(2), 127–140.

Botella, C., Serrano, B., Baños, R. M., & Garcia-Palacios, A. (2015). Virtual reality exposure-based therapy for the treatment of post-traumatic stress disorder: A review of its efficacy, the adequacy of the treatment protocol, and its acceptability. *Neuropsychiatric Disease and Treatment, 11*, 2533–2545.

Brewster, D. (1856). *The stereoscope; Its history, Theory and construction, with its application to the fine and useful arts and to education. Etc.* John Murray.

Carrozzino, M., & Bergamasco, M. (2010). Beyond virtual museums: Experiencing immersive virtual reality in real museums. *Journal of Cultural Heritage, 11*(4), 452–458.

Cha, M., Han, S., Lee, J., & Choi, B. (2012). A virtual reality based fire training simulator integrated with fire dynamics data. *Fire Safety Journal, 50*, 12–24.

Cheok, A. D., Tewell, J., Pradana, G. A., & Tsubouchi, K. (2013). Touch, Taste, and Smell: Multi-sensory Entertainment. In *International Conference on Advances in Computer Entertainment Technology* (pp. 516–518). Springer, Cham.

Crone, R. A. (1992). The history of stereoscopy. *Documenta Ophthalmologica, 81*(1), 1–16.

Dinh, H. Q., Walker, N., Hodges, L. F., Song, C., & Kobayashi, A. (1999). Evaluating the importance of multi-sensory input on memory and the sense of presence in virtual environments. In *Proceedings IEEE Virtual Reality* (Cat. No. 99CB36316) (pp. 222–228). IEEE, Houston, TX.

Emmelkamp, P. M., Krijn, M., Hulsbosch, A., De Vries, S., Schuemie, M. J., & van der Mast, C. A. (2002). Virtual reality treatment versus exposure in vivo: A comparative evaluation in acrophobia. *Behaviour Research and Therapy, 40*(5), 509–516.

Fisher, S. S., Wenzel, E. M., Coler, C., & McGreevy, M. W. (1988). Virtual interface environment workstations. In *Proceedings of the Human Factors Society Annual Meeting 32* (2), Los Angeles, CA: SAGE Publications: pp. 91–95.

Franklin Institute. (2019). *History of virtual reality*. Philadelphia: The Franklin Institute. Retrieved July 20, 2019, from https://www.fi.edu/virtual-reality/history-of-virtual-reality

Freeman, D., Reeve, S., Robinson, A., Ehlers, A., Clark, D., Spanlang, B., & Slater, M. (2017). Virtual reality in the assessment, understanding, and treatment of mental health disorders. *Psychological Medicine, 47*(14), 2393–2400.

Gaffney, P. (2010). *The force of the virtual: Deleuze, science, and philosophy*. University of Minnesota Press.

Garcia-Betances, R. I., Jiménez-Mixco, V., Arredondo, M. T., & Cabrera-Umpiérrez, M. F. (2015). Using virtual reality for cognitive training of the elderly. *American Journal of Alzheimer's Disease & Other Dementias, 30*(1), 49–54.
Heilig, M. L. (1962). *Sensorama simulator.* Google Patents.
Herrero, P., & De Antonio, A. (2005). Intelligent virtual agents keeping watch in the battlefield. *Virtual Reality, 8*(3), 185–193.
Hettinger, L. J., & Riccio, G. E. (1992). Visually induced motion sickness in virtual environments. *Presence: Teleoperators & Virtual Environments, 1*(3), 306–310.
Hoffman, H. G., Patterson, D. R., & Carrougher, G. J. (2000). Use of virtual reality for adjunctive treatment of adult burn pain during physical therapy: A controlled study. *The Clinical Journal of Pain, 16*(3), 244–250.
Hoffman, H. G., Chambers, G. T., Meyer, W. J., III, Arceneaux, L. L., Russell, W. J., Seibel, E. J., Richards, T. L., Sharar, S. R., & Patterson, D. R. (2011). Virtual reality as an adjunctive non-pharmacologic analgesic for acute burn pain during medical procedures. *Annals of Behavioral Medicine, 41*(2), 183–191.
IDC. (2018). *Worldwide quarterly augmented and virtual reality headset tracker.* International Data Corporation. https://www.idc.com/tracker/showproductinfo.jsp?prod_id=1501
Kennedy, R. S., Drexler, J., & Kennedy, R. C. (2010). Research in visually induced motion sickness. *Applied Ergonomics, 41*(4), 494–503.
Kinateder, M., Ronchi, E., Nilsson, D., Kobes, M., Müller, M., Pauli, P., & Mühlberger, A. (2014). Virtual reality for fire evacuation research. In *2014 federated conference on computer science and information systems* (pp. 313–321). IEEE, Warssaw.
Krueger, M. W. (1993). An easy entry artificial reality. In A. Wexelblat (Ed.), *Virtual reality: Applications and explorations* (pp. 147–161). Elsevier.
Krueger, M. W., & Wilson, S. (1985). VIDEOPLACE: a report from the artificial reality laboratory. *Leonardo, 18*(3), pp. 145–151.
Lele, A. (2013). Virtual reality and its military utility. *Journal of Ambient Intelligence and Humanized Computing, 4*(1), 17–26.
Lepouras, G., & Vassilakis, C. (2004). Virtual museums for all: Employing game technology for edutainment. *Virtual Reality, 8*(2), 96–106.
Liu, S. (2019). *Unit shipments of virtual reality (VR) devices worldwide from 2017 to 2019 (in millions), by vendor.* https://www.statista.com/statistics/671403/global-virtual-reality-device-shipments-by-vendor/ (Accessed 24/06/2020)
Mann, S. (2015). Phenomenal augmented reality: Advancing technology for the future of humanity. *IEEE Consumer Electronics Magazine, 4*(4), 92–97.
McGee, M. T. (2001). *Beyond Ballyhoo: Motion picture promotion and Gimmicks.* McFarland.

Milgram, P., Takemura, H., Utsumi, A., & Kishino, F. (1995). Augmented reality: A class of displays on the reality-virtuality continuum. *Telemanipulator and Telepresence Technologies, 2351*, 282–292.

Moulard-Leonard, V. (2008). *Bergson-Deleuze encounters: Transcendental experience and the thought of the virtual*. SUNY Press.

Olofsson, J. K., Niedenthal, S., Ehrndal, M., Zakrzewska, M., Wartel, A., & Larsson, M. (2017). Beyond smell-o-vision: Possibilities for smell-based digital media. *Simulation & Gaming, 48*(4), 455–479.

Orr, T. J., Mallet, L., & Margolis, K. A. (2009). Enhanced fire escape training for mine workers using virtual reality simulation. *Mining Engineering, 61*(11), 41.

Paul, S. (2009). Binaural recording technology: A historical review and possible future developments. *Acta Acustica United with Acustica, 95*(5), 767–788.

Portman, M. E., Natapov, A., & Fisher-Gewirtzman, D. (2015). To go where no man has gone before: Virtual reality in architecture, landscape architecture and environmental planning. *Computers, Environment and Urban Systems, 54*, 376–384.

Powers, M. B., & Emmelkamp, P. M. (2008). Virtual reality exposure therapy for anxiety disorders: A meta-analysis. *Journal of Anxiety Disorders, 22*(3), 561–569.

Pujol, L. (2004). Archaeology, museums and virtual reality. *Digithum, 6*, 1–9.

Reed, S., Kreylos, O., Hsi, S., Kellogg, L., Schladow, G., Yikilmaz, M., Segale, H., Silverman, J., Yalowitz, S., & Sato, E. (2014). *Shaping watersheds exhibit: An interactive, augmented reality sandbox for advancing earth science education*. AGU Fall Meeting Abstracts, abstract id. ED34A-01

Reger, G. M., Holloway, K. M., Candy, C., Rothbaum, B. O., Difede, J., Rizzo, A. A., & Gahm, G. A. (2011). Effectiveness of virtual reality exposure therapy for active duty soldiers in a military mental health clinic. *Journal of Traumatic Stress, 24*(1), 93–96.

Reznick, R. K., & MacRae, H. (2006). Teaching surgical skills—Changes in the wind. *New England Journal of Medicine, 355*(25), 2664–2669.

Rizzo, A., Pair, J., McNerney, P. J., Eastlund, E., Manson, B., Gratch, J., Hill, R., & Swartout, B. (2005). Development of a VR therapy application for Iraq war military personnel with PTSD. *Studies in Health Technology and Informatics, 111*, 407–413.

Rizzo, A., Buckwalter, J. G., John, B., Newman, B., Parsons, T., Kenny, P., & Williams, J. (2012). STRIVE: Stress resilience in virtual environments: A pre-deployment VR system for training emotional coping skills and assessing chronic and acute stress responses. *Studies in Health Technology and Informatics, 173*, 379–385.

Rothbaum, B. O., Hodges, L., Smith, S., Lee, J. H., & Price, L. (2000). A controlled study of virtual reality exposure therapy for the fear of flying. *Journal of Consulting and Clinical Psychology, 68*(6), 1020–1026.

Ryan, M.-L. (2001). *Narrative as virtual reality. Immersion and interactivity in literature*. Johns Hopkins University Press.
Sanchez-Vives, M. V., & Slater, M. (2005). From presence to consciousness through virtual reality. *Nature Reviews Neuroscience, 6*, 332–339.
Seymour, N. E., Gallagher, A. G., Roman, S. A., O'brien, M. K., Bansal, V. K., Andersen, D. K., & Satava, R. M. (2002). Virtual reality training improves operating room performance: Results of a randomized, double-blinded study. *Annals of Surgery, 236*(4), 458.
Sutherland, I. E. (1965/2002). The ultimate display. In R. Packer & K. Jordan (Eds.), *Multimedia: From Wagner to virtual reality*. WW Norton & Company.
Torick, E. (1998). Highlights in the history of multichannel sound. *Journal of the Audio Engineering Society, 46*(1/2), 27–31.
Wade, N., & Ono, H. (1985). The stereoscopic views of Wheatstone and Brewster. *Psychological Research, 47*(3), 125–133.
Wann, J., & Mon-Williams, M. (1996). What does virtual reality NEED?: Human factors issues in the design of three-dimensional computer environments. *International Journal of Human-Computer Studies, 44*(6), 829–847.
Willaert, W. I., Aggarwal, R., Van Herzeele, I., Cheshire, N. J., & Vermassen, F. E. (2012). Recent advancements in medical simulation: Patient-specific virtual reality simulation. *World Journal of Surgery, 36*(7), 1703–1712.

CHAPTER 2

The Ethical Dimensions of Virtual Reality

Abstract This chapter discusses the ways in which cultural and epistemic fears over virtual objects and environments emerged in the early 1980s as the fear of *covert* artificial reality was posed in literature and film. Though the promise of VR as a fully immersive alternative reality was let down by the lack of computing power to achieve such a vision, ethical concerns increasingly focused upon the way in which virtual objects have begun to bleed into our public life through phenomena such as 'deep fake' videos and chatbots, which when used maliciously threaten personal privacy, dignity, political security and public trust in the norms and institutions of democratic governance. The cghapter then examines the potentially negative behavioural impacts of VR upon human-agent and human-computer interactions (in terms of privacy, health, behaviour and well-being) and questions whether VR can be employed to stimulate prosocial and morally beneficial behaviours amongst users and their online and offline social networks.

Keywords Deep fakes • Covert virtual reality • Ethics • Privacy

M. Cotton, *Virtual Reality, Empathy and Ethics*,
https://doi.org/10.1007/978-3-030-72907-3_2

Introduction

Virtual reality (VR) is a component of a broader digital-cultural revolution that has gained pace since the late 1980s. Cross-border and cross-cultural digital communication and the growth in technologically mediated social relationships has evolved from the first dial-up bulletin boards, to the World Wide Web, to a system now dominated by global search engines and social media giants. These changes to communicative practices and the nature of social networks have led philosophers to question whether or not such technologies are of positive benefit to our health, well-being and sense of social connection, community and belonging. As Lévy (1998) notes, the technological mediation of social life via the internet is transforming the virtual into a type of collective intelligence, one that philosophers are concerned is slowly dehumanising our society. Thus, despite the relative advantages of VR for professional training, education and communication applications discussed in Chap. 1, there remain a number of serious personal and societal risks associated with its use.

Virtual Reality in the Popular Imagination

Since the early development of commercial VR technologies in the late 1980s and early 1990s, there have been ongoing cultural and epistemic fears that it presents a serious threat to human well-being, social cohesion and moral behaviour (Cogburn & Silcox, 2014; Beardon, 1992; Ford, 2001). As with many rapid technological advances, ethical concerns relate not only to the direct impacts of design, production and consumption (such as the environmental impacts or effects upon workers involved in the manufacturing process), but also upon the governance of technological artefacts and the way in which they are used. In short, the problem is not the technology per se, but rather who controls it and to what end?

Cultural fears about the governance of VR became a strong element of mainstream public understanding of the technology in the 1990s. Fears over the social control of technology were most clearly expressed through dystopian VR-themed creative outputs. Alongside concerns about the influence of cybernetics, the internet and artificial intelligence, VR has been the subject of numerous futuristic and dystopian science fiction books in the 1980s and 1990s—from William Gibson's *Neuromancer*, Tad William's *Otherland* series, or *Ready Player One* by Ernest Cline (which was made into a 2018 film directed by Steven Spielberg). The visual promise of VR also proved highly popular in film and television, in part because it allows film-makers the freedom to explore alternative realities as a means to generate dramatic tension. Films such as *Tron, The Lawnmower Man, Total Recall, Johnny Mnemonic, Existenz, The Matrix, Vanilla Sky, Enders Game* and the aforementioned *Ready Player One* (in chronological order) depict applications of VR that lead to commonly violent and destructive outcomes.

VR as a plot device used in cinema, literature and other forms of fiction differs substantially from the way in which it is actually used as a consumer or professional technology today. The reason for this is primarily technological. Since Krueger's (Krueger et al., n.d.; Krueger, 1993) concept of an artificial reality was first introduced, the actual design input and computing power necessary to create a functioning 'other world' has been severely limited. When the promise of a hyperrealistic alternative reality was deemed impossible given computing constraints in the 1990s, the technology fell out of fashion and became restricted to small-scale and specialised applications for business, architecture and training and so on. What differentiated current VR technologies from the dystopian realisation of VR on the silver screen is an ethical question of autonomy and consent. Contemporary VR technologies require the user to *actively* participate in the virtual environment that is created (and thus consent to do so). One must purchase or borrow a VR headset and controllers in order to experience it. In fiction, however, VR technologies are commonly used in a *covert* or involuntary manner. Films such as the Matrix, Vanilla Sky or Existenz display VR from the perspective of protagonists who cannot tell the difference between the virtual and the real. In other words, VR is used as a plot device to show how characters cannot fully perceive the nature of their own reality, or those that control it—that their participation in a virtual environment is through trickery, and the dramatic tension stems from not knowing who and what to trust. Fictionalised VR is presented as what I would term a *complete reality* which lies closer to Krueger's vision of an artificial reality: immersion is total and indistinguishable. VR is commonly presented within a science fiction trope: complete reality describes technological sophistication to the point at which the individual human agent is unaware of the distinction and thus there is no ontological separation between the virtual and the actual. Thus, philosophers and sociologists concerned with the longer-term societal impacts of VR have highlighted this ontological problem—as reality-altering technologies become increasingly sophisticated, computer-generated environments become difficult to distinguish from virtual ones, and so individuals need to make increasingly sophisticated judgements about what is real and what is not. This is an ethical problem because if individuals cannot tell the difference, they cannot give informed consent, and thus need protection by state, legislative and legal practices.

Covert Virtual Reality, Privacy and Data Security

Though a fully realised and indistinguishable technology-generated artificial or 'complete' reality remains science fiction, concerns over the risks of covert VR are not hypothetical observations about the outcome of an as yet unavailable futuristic technology. Of particular concern at the time of writing, is the way in which the virtual world is bleeding into the real world through computer-simulated realities that are increasingly difficult to distinguish from truth. So-called *deep fakes* are of particular concern. Deep fakes are manipulated videos that simulate the likeness of an individual through computer-generated video and manipulated audio of real people (usually politicians or other public figures such as celebrity spokespeople). In the video, the individual appears to say and do things that they never actually said or did. The technology of manipulating video is assisted by machine learning that copies the mannerisms and intonation of the person. Such technology has been used for the purposes of political satire. For example, in December 2020, in the United Kingdom the broadcaster Channel 4 showed a five-minute deep fake video of the Queen, voiced by the actor Debra Stephenson, in which she appeared to share her reflections on the year 2020, the departure of Prince Harry and Meghan Markle from active involvement in the Royal Household, and upon the Duke of York's alleged criminality and involvement in the scandal surrounding the alleged sex offender Jeffrey Epstein. The broadcast was intended to warn viewers of the dangers of fake news in this new technologically mediated deep fake context, with Channel 4 programmes director Ian Katz stating that the video is "a powerful reminder that we can no longer trust our own eyes" (cited in Blackall, 2020).

As the algorithms for copying human mannerisms become more sophisticated, social media sites such as YouTube™ will become infiltrated with evermore realistic depictions of key figures saying and doing things that they would not otherwise have done. Deep fakes create a representation, essentially a virtual person, that is resistant to detection from normal human senses of sight and hearing. Both the output of deep fake manipulation and the technologies to do the manipulation are rapidly diffused through social media networks, further eroding the truthfulness, legitimacy and public trust in digital content. Under such conditions, individuals and businesses face novel forms of exploitation, intimidation and personal sabotage from these technologies, with concomitant risks to democratic norms and potentially to national security (Chesney & Keats Citron, 2018).

There are other technologies potentially of concern which simulate human activity. So-called chatbots are one example. A chatbot is a piece of software which conducts a conversation either via auditory or textual methods. Like the deep fake, chatbots are designed to convincingly simulate human speech patterns, and the most advanced of these technologies allow the computer-simulated conversant to appear to behave like any other conversational partner. Chatbots are typically used in automated dialogue systems for information retrieval, such as in customer service, or query functions on websites. They vary in sophistication. Simpler chatbot systems scan for keywords within a typed query and then formulate a response which matches as many of the keywords as possible, or else scans a database of key terms and provides a pre-scripted response. Other more sophisticated systems employ natural language processing—becoming closer to a conversational artificial intelligence which can automate communication and make it appear to be more personalised to the user. One noticeable example was Microsoft's experiment with 'Tay', a combination of a chatbot and artificial intelligence programme. In March 2016 Tay caused controversy when the bot began to post inflammatory, derisive and offensive tweets to its twitter account shortly after launch. Microsoft claimed that the bot was attacked by trolls, and the system learned through its interactions, quickly becoming offensive based upon its interactions with others on the social media platform. The technology of automated conversation combined with machine learning creates what Neff and Nagy (2016) term *symbiotic agency*—the combination of individual expectations for the technology and the imagined affordances that are emerging in interaction between people and artificial intelligence, creates new forms of agency based upon new types of human-computer interaction.

Chatbot technologies are used primarily in the messaging applications as part of company websites, and they can also be used in banking systems or telephone-operated customer service. However, the new symbiotic agency of chatbots now extends to learning and play. The HelloBarbie™ toy has an internet-connected chatbot system developed by the company

ToyTalk™. ToyTalk developed the chatbot originally as a smartphone-based app to tell stories to children. Different characters can be interacted with, and the characters' behaviours are constrained by a set of rules that, in effect, emulate a particular character and thus produce a storyline. The HelloBarbie toy is the physical manifestation of a chatbot—it is a fashion doll that uses natural language processing technologies to construct a two-way conversation with the child owner. By using speech recognition and progressive learning features, the chatbot, like other forms of VR, helps to stimulate the sense of immersion and engagement in the play activity. The toy learns from the play history and this helps to tailor conversations and shape other interactive conversational elements such as storytelling or jokes.

Though chatbots are commonly used to save on the labour costs of customer service in information retrieval systems, there are potential malicious uses. Chatbots can be used as a form of interactive malware seeking to lure people not through threats or blackmail, but through seductive conversation; using psychological tricks and social engineering to target unprepared victims (Jonathan et al., 2009). They can be used to fill online messaging forums and chat rooms with spam or advertisements. They can also be used to entice people into revealing personal information such as addresses, contact details or bank details. Also, as with deep fakes, as chatbots become more sophisticated it becomes easier to craft seemingly real online identities for them. These seemingly human bots are sometimes referred to as 'sock puppets'. These can be employed to sway public opinion on a range of different political or social topics on social media platforms and chat rooms. Malicious bots can spread fake news in a way that improves the plausibility of the information provided. People are likely to be more willing to trust information they are given if it appears to come from a persuasive human agent during the normal course of a conversation.

Data breaches, the spread of misinformation and fake news created by virtual replicas and deep fakes is of growing public concern in an era of high-profile data security and public privacy controversies created by social media giants such as Facebook™. Facebook was found to have handed over the personally identifiable information of more than 87 million of its users to the firm Cambridge Analytica (CA). CA is a UK-based political consulting firm that is specialised in data-mining, so-called big data analysis and data brokerage, alongside strategic communication to political parties during electoral processes. CA was involved in 44 United States political races in 2014; then in 2015 advised Ted Cruz's 2015 presidential campaign, and then Donald Trump's 2016 successful presidential campaign. CA also provided data for the Leave.EU campaign during the United Kingdom's referendum on European Union membership.

Data was commonly harvested through online quizzes and games presented as third-party apps on Facebook; they also purchased from data brokers such as Acxiom and Nielson. CA used this mass of data to engage in what it termed *behavioural micro-targeting*—sorting high numbers of potential voters (estimated at 220 million Americans) into a series of behavioural and then tailoring specific political messages in support of the Trump campaign anchored in the psychological traits of each recipient based upon their profile. As Ward (2018) articulates, the messages played upon the hopes, fears and prejudices identified by the profile characteristics, and the message recipients may not have even been consciously aware that they held such views. CA claimed that the content produced using behavioural micro-targeting garnered 1.4 billion online impressions and led to tangible gains for the Trump campaign (ibid.). The power of big data analysis and targeted messaging was deeply influential in pushing the Trump campaign ahead of the rival Clinton campaign, and ultimately helped to win him the presidency through voter targeting in key districts of so-called swing states.

The ethical issues surrounding CA's role are twofold. First, is the manner in which data from Facebook users was gathered and assessed. Users that consented to give information to third-party apps such as games and quizzes were unaware that they were consenting for that data to be shared with CA for use in political campaigning. Second is the manner in which targeted messaging shapes the views and opinions of voters during the heightened emotional state of an election campaign. By paying for the services of CA, such an action is tantamount to buying political support, due to the efficacy of psychologically manipulative behavioural micro-targeting. CA's role in both the Trump and Leave.EU campaigns was politically controversial, and the company is subjected to ongoing criminal investigations in both the United States and United Kingdom. The CA scandal showed that established institutions—especially the mainstream media and political-party organisations—had already lost most of their power to shape public discourse, as big data analysis and new delivery vehicles for highly targeted data-driven fake news had overtaken their capacity to shape public opinion (hence voter intention) (Persily, 2017).

The outcome of the CA scandal is that public policymakers and public action campaigns are beginning to question the growth of academic research influenced by state security agendas (Laterza, 2018), are calling for tighter restriction and the adoption of comprehensive data privacy laws across national and international borders (Isaak & Hanna, 2018), and for research, development and regulation to catch up with malicious chatbot and deep fake algorithms that are becoming evermore difficult to differentiate from real conversation and political speech. The scandal also visibly highlighted the problems of data-mining in the context of global social media interactions whereby users freely give away private and personal information with few legal protections. The combination of targeted messaging on social media through data-mining and the exposure of voters to deep fakes that manipulate voter intention through dishonest and covert VR simulation is a powerful risk to democratic norms of free and fair elections, and deeper institutional trust. The risk is that individuals do not (and cannot) consent to being exposed to virtual realities—simulations of real people will appear authentic, and the victims are lacking protection by the law (Harris, 2018; Hasen, 2019). The deception that comes from expertly simulated environments without consent is a real moral threat that will grow as the technologies advance and are more widely dispersed across the internet. Certainly governance and accountability mechanisms to safeguard political campaigning through the verification of messages by spokespeople, technology to spot fakes and identify their origins, protection of individual reputations in the face of legal threats in defamation cases, and actions to shore up broader public trust in political and media institutions are going to remain significant challenges for political and legal institutions in the coming decades.

The Ethical Challenges of Virtual Reality

Covert VR presents real threat of harm to democratic norms, data privacy and public trust in political institutions. However, even when the technology is used with active consent, there are potential ethical pitfalls, and these have been subject to sustained philosophical evaluation. The ethics of VR are primarily related to the qualities of the virtual environments that are created, and of the behaviour of human agents within virtual worlds (Wilson, 1996; Beardon, 1992; Sherman & Judkins, 1992). The latter is important, because VR is increasingly understood in terms of *experiential ethics* rather than simply technological/machine-oriented ethics and regulatory systems (Steuer, 1994)—using VR influences the way in which human agents operate in both the virtual and physical-social real worlds. Identifying both the beneficial and malevolent influences upon human behaviour is therefore worthy of serious ethical reflection and empirical research.

There are situations whereby user interactions produce social behaviour and norms in virtual environments that are different to those that would normally emerge through social interaction in the physical world (Yee et al., 2007). What is needed, therefore, are new forms of collective social responsibility to effectively govern virtual ethical norms (Sharma et al., 2015). This raises questions about the moral dimensions of such changes—principally of *moral psychology*—the role that immersion and presence in virtual worlds alters the behaviours and actions of users, whether action in a virtual environment has a similar impact to action in a non-virtual environment, and whether there are spill-over effects from behaviours in virtual spaces into other forms of social interaction in non-virtual spaces. Certainly the promises and pitfalls of so-called *virtual communities*—spaces of public social interaction and deliberation between individuals who share common interests and common communicative practices but not the same physical space, are of growing ethical importance (Wilbur, 2013; Robins & Webster, 2003) as these spaces expand to cover not just gaming environments, but museums, learning platforms, corporate meetings or virtual religious services (Schroeder et al., 1998; Freina & Ott, 2015). Specifically, the creation of virtual environments allows freedom from real-world consequences to the individual, especially if the identity of users remains anonymous. Freedom from consequence leads technology ethicists and moral psychologists to question whether moral and immoral behaviour in VR is equivalent to that in everyday offline interaction (Brey, 1999), and also whether the actions of interactive non-player characters (NPC) and multiple-user virtual environments have moral weight (Ford, 2001). Moreover, as a question of moral psychology—does immoral action within virtual worlds influence behaviours in the 'real world'? Given the unlimited scope for designing new types of environments and encounters, what roles and responsibilities do designers of such environments have to ensure the upholding of moral norms? Thus, the development of VR and allied technologies such as computer-simulated virtual worlds in interactive Massive Multiplayer Online Role Playing Games (variously: World of Warcraft™, Guild Wars™, or Elder Scrolls Online™, to name a few currently played at the time of writing) or interactive worlds such as

Second Life™ pose similar or related ethical concerns (Wankel & Malleck, 2010).

Though debates over the ethical dimensions of VR emerged principally in the early 1990s (Steuer, 1994), the renewed commercial interest in the development of domestic motion-sensitive headsets for PCs and games consoles in the mid-to-late-2010s, including (at the time of writing) Facebook's Oculus™ platform, the HTC and Valve Vive ™ and Microsoft's Mixed Reality™ HoloLens™ technology and Sony's Morpheus™ platform, have presented a range of novel ethical issues around the risks and dangers that are foreseeable with the widespread use of VR. Much of the research on behavioural responses to VR and other interactive online spaces has focused upon the negative impacts of such environments, particularly gaming environments used by children and young adults. In Western countries, Rauterberg (2004) notes that discussion has predominantly focused upon violent game and media content and the effects upon behavioural development, social engagement, cognition and moral values; whereas in Japan, research has focused more upon intensive game usage and the impact on the intellectual and social development of children. The issue of violence is both politically and morally important. Physiological research of children gaming shows that participants who play violent games reported significantly higher states of anxiety and hostility (Arriaga et al., 2006), and so research linking game violence and players' expressions of aggression has been paramount within the literature (DeLisi et al., 2013). Meta-analysis of available psychological and physiological evidence shows a strong causal link between exposure to violent video games and increased aggressive behaviour, aggressive cognition, and aggressive affect and conversely decreased empathy and a decrease in prosocial behaviour (Anderson et al., 2010). Though the research focus has been upon the direct influence of violence in games upon violent behaviour and intentions of gamers, there are other social and contextual factors that mediate aggressive responses. For example, young people are often susceptible to messaging and manipulation through interactive environments (not least of which branding and product placement, see Mackay et al., 2009), the immersive nature of VR may therefore amplify the existing negative effects commonly experienced in other online platforms.

The negative neurological, behavioural and emotional effects on young people from increased reliance on screen-related recreational activity has made headline news (Wethington et al., 2013; Page et al., 2010), in particular due to reduced physical activity (and thus overall health and wellness due to increased risks from, e.g. chronic obesity, heart disease or type 2 diabetes). However, VR and AR can potentially alleviate these concerns. VR and AR such as the Pokémon Go™ game mentioned in Chap. 1 increase physical activity by encouraging users to roam and explore their environment, with concomitant positive impacts upon health (Rizzo et al., 2011). It must be noted also that studies of physical movement within gaming environments show that social engagement increases when body movement is afforded (Lindley et al., n.d.), and that gaming environments that promote *prosocial* (or altruistic) behaviours such as cooperation and empathy have corresponding effects on prosocial actions and behaviours in offline social environments (Gentile et al., 2009; Greitemeyer & Osswald, 2010). Such use also relates to corresponding improvements to mental health and well-being.

In addition to concerns raised about the effect of user actions within virtual environments or their immediate effects upon cognition and physiology, are the broader ethical issues raised by the wider sociocultural context of VR implementation. As numerous philosophers of technology have noted, technological change alters the broader moral landscape—by raising new ethical questions and proposing new answers, by changing the range of decisions and options available to moral decision makers and by altering human nature and social interactions (Harris, 2011; Luppicini, 2010; Verbeek, 2006, 2011). As Madary and Metzinger (2016) argue, technological innovation fundamentally changes the objective world, objective changes are subjectively perceived, which in turn may shift value judgements, potentially changing our understanding of 'conscious experience', 'selfhood', 'authenticity', or 'realness', and thus transforming our social interactions with others. Potential problems arise through the capacity of virtual spaces to augment human behaviour, and whether VR amplifies such changes. For example, Suler (2004) notes that whilst online, people commonly self-disclose, act out and engage more intensely than when they interact in offline social environments. Such an online *disinhibition effect* is rooted in (amongst other things) dissociative anonymity, a sense of social invisibility, and the minimisation of authority such that these cause shifts to the sense of self within such environments. Concerns are raised therefore over the potential of virtual environments to normalise changes in the sense of self over time, such that behavioural impacts spill over into broader offline social interactions. This is an ethical issue because certain online and virtual environments have become sites of harassment, doxing (publishing private/personal information online), bullying, grooming, simulated sexual harassment or assault, alongside any other forms of adverse social interaction that cause harm (Strikwerda, 2015). Thus, poorly regulated VR systems present significant threats to informational privacy, threats to physical privacy and associational privacy (O'Brolcháin et al., 2016), which require social regulation and stronger governance and control through legislation and law enforcement.

Within a virtual environment, immersion and presence can increase (for discussion see Farrow & Iacovides, 2014). Simulated harm that occurs to avatars (representatives of the individual in an online form) therefore creates virtual victims of online deviant behaviour. This can be distressing for participants when their avatars are harmed by other participants' malicious actions, yet as Wolfendale (2007) notes, there is a tendency in the literature on this topic to dismiss such distress as evidence of too great an involvement in and identification with the online character, rather than to treat such threats as being of legal and ethical concern. By contrast, Wolfendale (ibid.) asserts that avatar attachment is expressive of identity and self-conception and should therefore be accorded the moral significance we give to real-life attachments that play a similar role (such as personal, cultural or religious artefacts). Schechtman (2012) similarly suggests that understanding this continuity of online and offline selves through a narrative approach to identity is of philosophical and social scientific relevance—there is a continuum of identities such that threats to one influence the other. Notions of embodiment, that is, the extent to which a user feels the impacts upon virtual avatars within gaming or other online environments, is controversial (Farrow & Iacovides, 2014). However, within VR, avatars are more closely linked to the physiology of the individual through enhanced sensory engagement. This greater depth of sensory stimulation amplifies the embodiment effect (Kilteni et al., 2012). The possibility of real and permanent psychological harm linked to the avatar is thus exacerbated in unregulated virtual environments. Moreover, negative impacts experienced within virtual environments are not just immediate virtual threats, but also potentially lead to offline crime, intimidation and the potential normalisation of sexist, racist, homophobic and transphobic sentiments, thus generating a more generalised sense of social damage. As Spiegel (2018) argues, the potential threats to mental health, personal autonomy and personal privacy that VR environments present are a matter for urgent legislative and policy response.

The Ethical Benefits of Virtual Reality

Though the potential harms to the individual and to broader social norms are of significant philosophical concern, there are numerous positive applications of the technology which promote broader prosocial behavioural norms and positive impacts. As mentioned in the previous chapter, VR has many applications beyond violent (and non-violent) gaming, and these applications have inherent benefits to broader society. To briefly recap: VR has been highlighted as a tool for training (including military and other dangerous training scenarios) (Bhagat et al., 2016; Orr et al., 2009), product prototyping and business communication (Biocca & Levy, 2013; Mujber et al., 2004; Aromaa & Väänänen, 2016), science and cultural heritage communication (Pujol, 2004), education (Hedberg & Alexander, 1994; Schwienhorst, 2002; Freina & Ott, 2015; Seidel & Chatelier, 2013), medical training (Seymour et al., 2002) and therapeutic interventions (e.g. for the treatment of anxiety or social phobias, see Klinger et al., 2005), or in rehabilitation (Schultheis & Rizzo, 2001). Psychotherapists, in particular, have been interested in the technology's capacity to simulate social environments which would have positive and therapeutic value, for example in treating social phobia, fear of intimacy, sexual aversion, exploration of memory, expressing emotions, or increasing empathy (Glantz et al., 1996). Though the risks of unregulated virtual environments remain, they can also create interactions that promote positive psychological outcomes for users—they can be safe interactive spaces for engaging in therapeutic work, building social confidence or overcoming fears. Aside from beneficial therapeutic environments, we can also understand VR as a form of communicative media—the technology is increasingly used in news broadcast, in documentary film-making (Kaplan & Haenlein, 2009) or as an artistic medium (Bates, 1992), and these elements are discussed further in Chap. 5.

The social benefits of greater immersion in learning, therapy, contemporary political events, artistic expression and the narrative framing of people's stories is significant. VR can bring a closeness to interactions that other forms of digital communication can't, helping to facilitate deeper cultural connection through a heightened sense of immersion (Palmer, 1995). Since the 1970s, sociologists and human geographers have highlighted the issue of time-space compression between places, things and people (Massey, 1992). Rapid advances in mass transit, air travel, broadcast media and high-speed telecommunications have created a so-called shrinking world. Time-space compression is the experience of different places growing closer together through the connection of supply chains, transport and communication networks. Time-space compression is commonly understood in terms of travel time and cost—that global capitalism is dependent upon rapidly changing commodity markets, capital accumulation and a just-in-time model of production facilitated by rapid changes in transportation and telecommunications infrastructure (Harvey, 1999). The world shrinks because it is easier and quicker to move flows of goods and capital. However, though a sense of time and space is compressed in a new global digital economy, this is often accompanied by a growing sense of cultural disintegration, personal isolation and neo-tribalism in the face of corporate globalisation and mass immigration (see, e.g. Magda, 2001). Communities that had a sense of cohesion and integration bounded by geography, shared cultural symbols, religious practices and common language are now thrown open. Moreover, the growing power of internet-enabled social media creates new communicative networks across borders and cultures.

The digital revolution has facilitated multiculturalism, allowing greater communicative avenues between people of different ages, races, gender identities, and cultural and religious beliefs. And yet since the early 2000s there has been concern that increased use of internet-enabled technologies will negatively affect social communication (Anderson & Tracey, 2001). Indeed, there is evidence of new forms of communicative tribalism emerging (as discussed in relation to the Cambridge Analytica scandal discussed above). Internet forums that promote polarised political and cultural ideologies encourage the sharing and reinforcement of existing beliefs. This creates numerous social effects such as the echo-chamber effect or social bubbles that decrease the capacity for empathy between the different tribes. Yet one of the primary problems of internet-enabled communication is anonymity. Users can hide behind avatars or anonymous sock puppets in order to troll others: baiting arguments, or inciting hatred without consequence to themselves. The dehumanisation that occurs through digital interaction could be lessened through VR, because the feedback between interacting agents is both more personal (it is possible for individuals to meet in something akin to a physical encounter) and more visceral (agents can experience immersive sensory feedback, and potentially gauge intonation and body language from others). As discussed in Chap. 5 VR facilitates individuals' experience of narrative from multiple perspectives, even in the bodies of other agents. Journalists and artists hope that this will stimulate empathy amongst users—to better understand the embodied experiences of people from different cultural, ethnic, religious and political backgrounds, and thus improve the *moral* quality of online engagements between different social groups.

The question at the heart of this book, concerns the relationship between VR, social engagement and empathy—whether the positive outcomes experienced by users of VR in mental health settings, educational environments, training and science communication that come from great immersion, embodiment and digital interaction, can be harnessed to promote ethical reflection, decision-making and broader prosocial benefits for both online and offline social interactions. In the following chapters, VR is assessed in terms of its potential as an *ethical tool* (Weston, 2000) to facilitate personal reflection on complex ethical issues and positively influence decision-making outcomes to users.

References

Anderson, B., & Tracey, K. (2001). Digital living: The impact (or otherwise) of the Internet on everyday life. *American Behavioral Scientist, 45*(3), 456–475.

Anderson, C. A., Shibuya, A., Ihori, N., Swing, E. L., Bushman, B., Sakamoto, A., Rothstein, H. R., & Saleem, M. (2010). Violent video game effects on aggression, empathy, and prosocial behavior in Eastern and Western countries: A meta-analytic review. *Psychological Bulletin, 136*(2), 151–173.

Aromaa, S., & Väänänen, K. (2016). Suitability of virtual prototypes to support human factors/ergonomics evaluation during the design. *Applied Ergonomics, 56*, 11–18.

Arriaga, P., Esteves, F., Carneiro, P., & Monteiro, M. B. (2006). Violent computer games and their effects on state hostility and physiological arousal. *Aggressive Behavior: Official Journal of the International Society for Research on Aggression, 32*(2), 146–158.

Bates, J. (1992). Virtual reality, art, and entertainment. *Presence: Teleoperators & Virtual Environments, 1*(1), 133–138.

Beardon, C. (1992). The ethics of virtual reality. *Intelligent Tutoring Media, 3*(1), 23–28.

Bhagat, K. K., Liou, W.-K., & Chang, C.-Y. (2016). A cost-effective interactive 3D virtual reality system applied to military live firing training. *Virtual Reality, 20*(2), 127–140.

Biocca, F., & Levy, M. R. (2013). *Communication in the age of virtual reality*. Routledge.

Blackall, M. (2020). Channel 4 under fire for deepfake Queen's Christmas message *The Guardian*, Available: Guardian Media Group. https://www.theguardian.com/technology/2020/dec/24/channel-4-under-fire-for-deepfake-queen-christmas-message

Brey, P. (1999). The ethics of representation and action in virtual reality. *Ethics and Information Technology, 1*(1), 5–14.

Chesney, R., & Keats Citron, D. (2018). Deep fakes: A looming challenge for privacy, democracy, and national security. In 107 California Law Review (2019), F.U.o.T.L., Public Law Research Paper No. 692; U of Maryland Legal Studies Research Paper No. 2018–21. (ed.).

Cogburn, J., & Silcox, M. (2014). Against brain-in-a-vatism: On the value of virtual reality. *Philosophy & Technology, 27*(4), 561–579.

DeLisi, M., Vaughn, M. G., Gentile, D. A., Anderson, C. A., & Shook, J. J. (2013). Violent video games, delinquency, and youth violence: New evidence. *Youth Violence and Juvenile Justice, 11*(2), 132–142.

Farrow, R., & Iacovides, I. (2014). Gaming and the limits of digital embodiment. *Philosophy & Technology, 27*(2), 221–233.

Ford, P. J. (2001). A further analysis of the ethics of representation in virtual reality: Multi-user environments. *Ethics and Information Technology, 3*(2), 113–121.

Freina, L., & Ott, M. (2015). A literature review on immersive virtual reality in education: State of the art and perspectives. *eLearning & Software for Education, 1*.

Gentile, D. A., Anderson, C. A., Yukawa, S., Ihori, N., Saleem, M., Ming, L. K., Shibuya, A., Liau, A. K., Khoo, A., & Bushman, B. J. (2009). The effects of prosocial video games on prosocial behaviors: International evidence from correlational, longitudinal, and experimental studies. *Personality and Social Psychology Bulletin, 35*(6), 752–763.

Glantz, K., Durlach, N. I., Barnett, R. C., & Aviles, W. A. (1996). Virtual reality (VR) for psychotherapy: From the physical to the social environment. *Psychotherapy: Theory, Research, Practice, Training, 33*(3), 464.

Greitemeyer, T., & Osswald, S. (2010). Effects of prosocial video games on prosocial behavior. *Journal of Personality and Social Psychology, 98*(2), 211–221.

Harris, S. (2011). *The moral landscape: How science can determine human values.* Simon and Schuster.

Harris, D. (2018). Deepfakes: False pornography is here and the law cannot protect you. *Duke Law & Technology Review, 17*(1), 99–127.

Harvey, D. (1999). Time-space compression and the postmodern condition. *Modernity: Critical Concepts, 4*, 98–118.

Hasen, R. L. (2019). *Deep fakes, bots, and siloed justices: American election law in a post-truth world*, Irvine CA: UC Irvine School of Law Research Paper No. 2019–36.

Hedberg, J., & Alexander, S. (1994). Virtual reality in education: Defining researchable issues. *Educational Media International, 31*(4), 214–220.

Isaak, J., & Hanna, M. J. (2018). User data privacy: Facebook, Cambridge Analytica, and privacy protection. *Computer, 51*(8), 56–59.

Jonathan, P. J. Y., Fung, C. C., & Wong, K. W. (2009). Devious chatbots-interactive malware with a plot. In J.-H. Kim, S. S. Ge, P. Vadakkepat, J. Norbert, A. A. Manum, K. Sadasivan Puthusserypady, U. Rückert, J. Sitte, U. Witkowski, R. Nakatsu, T. Braun, J. Baltes, J. R. Anderson, C.-C. Wong, I. Verner, & D. Ahlgren (Eds.), *Progress in robotics. FIRA 2009. Communications in computer and information science, vol 44* (pp. 110–118). Springer.

Kaplan, A. M., & Haenlein, M. (2009). The fairyland of second life: Virtual social worlds and how to use them. *Business Horizons, 52*(6), 563–572.

Kilteni, K., Groten, R., & Slater, M. (2012). The sense of embodiment in virtual reality. *Presence: Teleoperators and Virtual Environments, 21*(4), 373–387.

Klinger, E., Bouchard, S., Légeron, P., Roy, S., Lauer, F., Chemin, I., & Nugues, P. (2005). Virtual reality therapy versus cognitive behavior therapy for social phobia: A preliminary controlled study. *Cyberpsychology & Behavior, 8*(1), 76–88.

Krueger, M. W. (1993). An easy entry artificial reality. In A. Wexelblat (Ed.), *Virtual reality: Applications and explorations* (pp. 147–161). Elsevier.

Krueger, M. W., Gionfriddo, T., & Hinrichsen, K. (n.d.). VIDEOPLACE—An artificial reality. *ACM SIGCHI Bulletin*: ACM, 35–40.

Laterza, V. (2018). Cambridge Analytica, independent research and the national interest. *Anthropology Today, 34*(3), 1–2.

Lévy, P. (1998). *Becoming virtual: Reality in the digital age*. Plenum Press.

Lindley, S. E., Le Couteur, J., & Berthouze, N. L. (n.d.). Stirring up experience through movement in game play: Effects on engagement and social behaviour. *Proceedings of the SIGCHI Conference on Human Factors in Computing Systems*: ACM, 511–514.

Luppicini, R. (Ed.). (2010). *Technoethics and the evolving knowledge society: Ethical issues in technological design, research, development, and innovation*. IGI Global.

Mackay, T., Ewing, M., Newton, F., & Windisch, L. (2009). The effect of product placement in computer games on brand attitude and recall. *International Journal of Advertising, 28*(3), 423–438.

Madary, M., & Metzinger, T. K. (2016). Real virtuality: A code of ethical conduct. recommendations for good scientific practice and the consumers of VR-technology. *Frontiers in Robotics and AI, 3*.

Magda, R. M. R. (2001). Transmodernity, neotribalism and postpolitics. *Interlitteraria, 6*(6), 9–18.

Massey, D. (1992). Politics and space/time. *New Left Review, 196*, 65–84.

Mujber, T. S., Szecsi, T., & Hashmi, M. S. (2004). Virtual reality applications in manufacturing process simulation. *Journal of Materials Processing Technology, 155*, 1834–1838.

Neff, G., & Nagy, P. (2016). Automation, algorithms, and politics| talking to Bots: Symbiotic agency and the case of Tay. *International Journal of Communication, 10*, 4915–4931.
O'Brolcháin, F., Jacquemard, T., Monaghan, D., O'Connor, N., Novitzky, P., & Gordijn, B. (2016). The convergence of virtual reality and social networks: Threats to privacy and autonomy. *Science and Engineering Ethics, 22*(1), 1–29.
Orr, T. J., Mallet, L., & Margolis, K. A. (2009). Enhanced fire escape training for mine workers using virtual reality simulation. *Mining Engineering, 61*(11), 41.
Page, A. S., Cooper, A. R., Griew, P., & Jago, R. (2010). Children's screen viewing is related to psychological difficulties irrespective of physical activity. *Pediatrics, 126*(5), e1011–e1017.
Palmer, M. T. (1995). Interpersonal communication and virtual reality: Mediating interpersonal relationships. In F. Biocca & M. R. Levy (Eds.), *Communication in the age of virtual reality* (pp. 277–299). Laurence Erlbaum and Associates Ltd.
Persily, N. (2017). The 2016 US Election: Can democracy survive the internet? *Journal of Democracy, 28*(2), 63–76.
Pujol, L. (2004). Archaeology, museums and virtual reality. *Digithum, 6*, 1–9.
Rauterberg, M. (2004). *Positive effects of entertainment technology on human behaviour. Building the information society*. Springer, pp. 51–58.
Rizzo, A. S., Lange, B., Suma, E. A., & Bolas, M. (2011). Virtual reality and interactive digital game technology: New tools to address obesity and diabetes. *Journal of Diabetes Science and Technology, 5*(2), 256–264.
Robins, K., & Webster, F. (2003). *Times of the technoculture: From the information society to the virtual life*. Routledge.
Schechtman, M. (2012). The story of my (second) life: Virtual worlds and narrative identity. *Philosophy & Technology, 25*(3), 329–343.
Schroeder, R., Heather, N., & Lee, R. M. (1998). The sacred and the virtual: Religion in multi-user virtual reality. *Journal of Computer-Mediated Communication, 4*(2). JCMC425, https://doi.org/10.1111/j.1083-6101.1998.tb00092.x
Schultheis, M. T., & Rizzo, A. A. (2001). The application of virtual reality technology in rehabilitation. *Rehabilitation Psychology, 46*(3), 296.
Schwienhorst, K. (2002). Why virtual, why environments? Implementing virtual reality concepts in computer-assisted language learning. *Simulation & Gaming, 33*(2), 196–209.
Seidel, R. J., & Chatelier, P. R. (2013). *Virtual reality, training's future?: Perspectives on virtual reality and related emerging technologies*. Springer Science & Business Media.
Seymour, N. E., Gallagher, A. G., Roman, S. A., O'brien, M. K., Bansal, V. K., Andersen, D. K., & Satava, R. M. (2002). Virtual reality training improves operating room performance: Results of a randomized, double-blinded study. *Annals of Surgery, 236*(4), 458.

Sharma, S., Lomash, H., & Bawa, S. (2015). Who regulates ethics in the virtual world? *Science and Engineering Ethics, 21*(1), 19–28.
Sherman, B., & Judkins, P. (1992). *Glimpses of heaven, visions of hell: Virtual reality and its implications*. Hodder & Stoughton.
Spiegel, J. S. (2018). The ethics of virtual reality technology: Social hazards and public policy recommendations. *Science and Engineering Ethics, 24*(5), 1537–1550.
Steuer, J. (1994). Defining virtual reality: Dimensions determining telepresence. *Journal of Communication, 42*(4), 73–93.
Strikwerda, L. (2015). Present and future instances of virtual rape in light of three categories of legal philosophical theories on rape. *Philosophy & Technology, 28*(4), 491–510.
Suler, J. (2004). The online disinhibition effect. *Cyberpsychology & Behavior, 7*(3), 321–326.
Verbeek, P.-P. (2006). Materializing morality: Design ethics and technological mediation. *Science, Technology, & Human Values, 31*(3), 361–380.
Verbeek, P.-P. (2011). *Moralizing technology: Understanding and designing the morality of things*. University of Chicago Press.
Wankel, C., & Malleck, S. (2010). Exploring new ethical issues in the virtual worlds of the twenty-first century. In C. Wankel & S. Malleck (Eds.), *Emerging ethical issues of life in virtual worlds* (pp. 1–14). Information Age Publishing.
Ward, K. (2018). Social networks, the 2016 US presidential election, and Kantian ethics: Applying the categorical imperative to Cambridge Analytica's behavioral microtargeting. *Journal of Media Ethics, 33*(3), 133–148.
Weston, A. (2000). *A 21st century ethical toolbox*. Oxford University Press.
Wethington, H., Pan, L., & Sherry, B. (2013). The association of screen time, television in the bedroom, and obesity among school-aged youth: 2007 National Survey of Children's Health. *Journal of School Health, 83*(8), 573–581.
Wilbur, S. P. (2013). An archaeology of cyberspaces: Virtuality, community, identity. In D. Porter (Ed.), *Internet culture* (pp. 5–22). Routledge.
Wilson, J. R. (1996). Effects of participating in virtual environments a review of current knowledge. *Safety Science, 23*(1), 39–51.
Wolfendale, J. (2007). My avatar, my self: Virtual harm and attachment. *Ethics and Information Technology, 9*(2), 111–119.
Yee, N., Bailenson, J. N., Urbanek, M., Chang, F., & Merget, D. (2007). The unbearable likeness of being digital: The persistence of nonverbal social norms in online virtual environments. *CyberPsychology & Behavior, 10*(1), 115–121.

CHAPTER 3

Technology Governance and Ethics

Abstract This chapter has three components. First is the discussion of technology governance and its relationship to broader public deliberation upon the ethical impacts of complex socio-technical systems. Second is a discussion of the relationship between the social control of technology through deliberative democratic means, and normative, empirical and applied ethics. Third is an assessment of the concept of 'ethical tools'—practical decision-support methods or heuristic devices to aid (rather than replace) processes of ethical reflection and judgement without recourse to normative theory. Finally, the potential of VR to act as an 'ethical tool' to facilitate judgement is mooted, and a conceptual framework developed in Chap. 4.

Keywords Ethical tools • Social control of technology • Participatory-deliberative methods

Introduction

As discussed in Chaps. 1 and 2, VR is potentially beneficial in a range of applications, from education, training, art, design and therapy. Yet it sparks a number of significant ethical dilemmas, particularly when virtual objects have unforeseen or unwanted influence upon human behaviours. There are many routes to ethical controversy in virtual environments. Whether through voluntary involvement in online gaming or through exposure to

M. Cotton, *Virtual Reality, Empathy and Ethics*,
https://doi.org/10.1007/978-3-030-72907-3_3

political manipulation through deep fake videos, the development and application of the technologies has broader impacts to social life, governance and institutional arrangements that go beyond the impacts to the individual user. The development of innovative technologies (whether in the form of VR, robotics, autonomous systems, big data harvesting, smart appliances or artificial intelligence and so on), is often 'stimulated' by government institutions. Technological artefacts, systems and processes are championed by government actors as a basic societal good. The technological optimism of government is a political choice. It is driven by what is sometimes defined as *competitive nationalism* (Law & Mooney, 2012)—whereby the declining manufacturing capacity and rise in service sector economic activity is perceived as an economic and political risk in European and North American economies relative to emergent economies that have far greater capacity in both commodities markets and high-tech manufacturing. The political solution to the problem of competitive nationalism is a type of 'boosterism'—usually a combination of 'push' and 'pull' policies (Nemet, 2009) to improve the development and uptake of technology within a broader industrial strategy. We can simplify here that a *push* strategy is used to incentivise innovation, research and technology development, and a *pull* strategy to help turn innovation into marketable products, to build manufacturing capacity, global competitiveness and to generate consumer interest.

Within an overarching neoliberal policy framework, innovation, technological development and marketisation are construed as key pathways to prosperity. Prosperity therefore roughly corresponds in the political imagination to a general improvement in overall national welfare. As Chap. 1 shows, the uptake of VR and the growing number of applications result in a rapid expansion of markets and user territories. The underlying ethical norm that drives the pro-innovation policy discourse is one whereby technology is beneficial so long as profit is made, jobs are produced and basic regulatory social and environmental standards are upheld. As Stirling et al. (2018) note, science and technology have become deeply entangled in broader concepts of social progress, such that most innovation is imagined by policymakers and business leaders as synonymous with the public good. However, despite the undoubted benefits to public health, education, wealth creation and social cohesion from numerous technological advances, the consequences of research and innovation are not always desirable, and unintended consequences can produce negative social outcomes as discussed in Chap. 2. Moreover, when benefits do emerge, they do not do so automatically, and they are not always evenly distributed across society. Therefore, good *technological governance* not only requires finding a balance between social gains and inherent risks from technological developments but must also be founded upon deeper questions over the nature and direction in which technologically mediated social progress is 'pushed' by public and private interests.

It is an established norm within the literatures on technology governance that innovation should involve some form of *social control*—usually through one or more types of strong democratic oversight of innovation development and implementation (Sclove, 1995; Kleinman, 2000). This is necessary due to the twin problems framed in Collingridge's dilemma (Collingridge, 1980). The first—what Collingridge describes as an *information problem*—is that both the positive and negative consequences of a technology are difficult to predict from the point of inception and are only fully understood when the technology is extensively developed and widely used. This leads to a second *power problem*—the act of controlling or changing a technology becomes increasingly difficult as it becomes widely adopted and then entrenched within social life. In essence, innovation can create a 'letting the genie out of the bottle' problem. The solution to this dilemma is analogous to the *precautionary principle.* Some attempt at foresight over the positive and negative impacts of technology futures is needed to solve the information problem, and then the adjustment of policy and legal/regulatory controls to anticipate negative impacts is necessary to resolve the power problem. Consequently, when technologies have a considerable impact upon the social life of users (and, by extension, non-users), social scientists and philosophers of technology have argued that ethical reflection and democratic accountability should be part of discussions over design, implementation, uptake, use, decommissioning and disposal of such technologies 'upstream' of their release into the market so that social control over the technology can be established (Cotton, 2010; Mohr, 2007; Wilsdon & Willis, 2004; Devon, 2004; Keulartz et al., 2004; Rip et al., 1995; Rip & Kemp, 1998; Cotton, 2014b).

The underlying ethos of this trend towards upstream social control of technology (SCOT) is a recognition that technology decisions have inherent political, sociocultural and ethical components. Technological programmes are understood as *socio-technical,* not merely technical—they are complex systems constructed from interrelated human and non-human elements. The stability of a social system of rules and norms is underpinned by technological systems that facilitate those norms, and conversely the technological systems that are developed, implemented and maintained are shaped by broader technological cultures, political decisions, user experiences, ethical and aesthetic values. The dialectical relationship between society and technology is therefore captured in this notion of a socio-technical system.

If we take VR to be a socio-technical system, then we assert that in order to capture the broader social benefits of the technology, we must first address the interests and values of heterogeneous stakeholders in ways that cannot be resolved using expert judgement, quantitative tools such as economic forecasting, risk assessment and cost-benefit analysis alone (Beierle, 2002; Wynne, 1996; Lidskog, 2008; Maranta et al., 2003). Engaging with a broad array of stakeholders (including non-expert citizens), will allow stronger evaluation of perceptions, practices, ethical values and social impacts within the design process. From a strategic perspective, this may elicit important information that would otherwise be overlooked in a purely technical analysis (Beierle, 1999; Fiorino, 1990), and so a *participatory-deliberative* approach that brings together diverse voices in a process of discussion, reflection and decision-making, can often work to improve the fairness and social robustness of technology governance outcomes (Nowotny, 2003). There has been a strong push within academic and policy circles over the last three decades towards what is commonly referred to as a *participatory-deliberative turn*—that democratic legitimacy is construed in terms of the direct capacity of citizens to be involved in decisions over technology governance, rather than being represented by an elected third-party through aggregate voting mechanisms at arm's length from the socio-technical system in question. In many democratic countries, the SCOT is through formal processes of technology assessment. Its aim is to focus social preferences, develop forecasts towards both the expected and unexpected social and environmental consequences of technological initiatives, and to strengthen long-range social planning in public decision-making processes (Wenk, 1975; Van Den Ende et al., 1998).

Technology assessment as a formal process involves expert independent review bodies that provide advice to governing bodies and thus ensure that conditions such as a public and environmental safety are met. However, although there are still some functioning independent technology advisory bodies that can oversee such programmes of engagement (such as the Parliamentary Office of Science and Technology or POST, in the United Kingdom), in many democratic nations such technology assessment bodies have closed or diminished (such as the United States' Office of Technology Assessment), meaning that governance of technology is increasingly left to laissez-faire market conditions, and controlled through light touch regulation. VR commonly classified as a consumer electronics product for gaming is one such example. There is therefore a contradiction within the rhetoric of technology politics—there is increased emphasis upon social responsibility and engagement in science and technology, yet this is coupled with simultaneous deregulation and boosterism that diminishes the capacity of citizens to seize the SCOT. Given the potential social harms from unrestricted and unregulated virtual environments discussed in Chap. 2, a laissez-faire market for access to virtual worlds is ethically problematic.

The drive to engage the public directly in technology governance emerged in response to successive failures in innovation policy. For example, public backlash against genetically modified organisms (Murphy et al., 2006), nano-technologies (Sheetz et al., 2005; Pidgeon & Rogers-Hayden, 2007), nuclear technologies (Konrad et al., 2018; Augustine, 2018; Cotton, 2017), or hydraulic fracturing for oil and gas (Whitton et al., 2017), are seen as failures of SCOT. When scientific and technical authorities impose technical rationality onto a fundamentally ethical problem, social movements of opposition will often mobilise to redress the value conflict that emerges. As a result, both national and transnational technology governance bodies have expressed commitments towards *responsible research and innovation* (RRI), in order to address the problem of stakeholder conflict. The ethos of RRI is to move from a production and output-focused engineering approach, towards a human-centred engineering approach—replacing emphasis upon product quality, efficiency, cost reduction, and precision, with a focus upon broader social and environmental benefits and values (Owen et al., 2012; Taebi et al., 2014). RRI has since become embedded within funding platforms for research grants and other innovation pathways under European Commission programmes; indeed they summarise the concept as follows (European Commission, 2018):

> an approach that anticipates and assesses potential implications and societal expectations with regard to research and innovation, with the aim to foster the design of inclusive and sustainable research and innovation … that societal actors (researchers, citizens, policy makers, business, third sector organisations, etc.) work together during the whole research and innovation

> process in order to better align both the process and its outcomes with the values, needs and expectations of society. ... a package that includes multi-actor and public engagement in research and innovation.

It is notable, however, that formal processes of RRI assessment are commonly applied to those technologies that have *anticipated* environmental or social harms associated with them (e.g. nano-technologies, nuclear fusion) (Healy, 2012), or else will spur deeper value conflicts amongst specific stakeholder groups (such as stem cell research, xeno-transplantation or gene therapies). These innovations can be categorised as socially and ethically contentions technologies (SECTs) (Cotton, 2014b). For SECTs, processes of technology governance often involve upstream engagement combined with formal legal and policy responses. For example, in the UK the GM Nation? public consultation on transgenic crops (Horlick-Jones et al., 2006), the Committee on Radioactive Waste Management's evaluation of nuclear waste disposal technologies (Blowers & Sundqvist, 2010; Cotton, 2017), or the Climate Assembly UK (Climate Assembly UK, 2020) involved complex processes of public engagement on ethical and value questions; the outcomes of which fed directly into shaping government policy. These processes are inherently precautionary, 'upstream' of technology implementation or policy implementation (Cotton, 2010; Wilsdon & Willis, 2004), and hence consonant with an RRI approach. Yet for VR—associated primarily with consumer gaming technology and entertainment—such processes of strong democratic SCOT are absent. The technology is viewed as benign because it is equated with entertainment. Under such circumstances, governments favour market solutions to technology development and uptake, with broader social impacts hidden from public scrutiny. Yet given the *diffuse* social harms from VR to create new unregulated and uncontrolled online social environments discussed in Chap. 2, it behoves government authorities to develop 'catch up' legislative responses to protect consumer welfare, given the rapid proliferation of virtual technologies across a range of overt and covert online applications. This doesn't mean an outright *ban* upon the technology, but as discussed above, greater SCOT—taking steps to maximise benefits and minimise societal harms, whilst focussing on broader issues of justice of access and governance of online and offline behaviours associated with virtual environments—is warranted given the risks at play.

Stakeholder Values and Ethical Assessment

The notion that technology requires social control through the participation and deliberation of multiple stakeholder actors has two implications for technology ethics. The first is a fundamental challenge to the idea that technologies are agnostic of value considerations, or are *ethically neutral* (Alcorn, 2001). Ethical neutrality implies that it is the cognitive process of individual moral judgement that controls how technologies are used and the technology exerts no direct influence upon that process. By extension, therefore, engineers, designers and other technical specialists are absolved from making ethical judgements about the artefacts that they create. This challenge to the ethical neutrality of artefacts has now become a core part of science and engineering ethics—that the ethical responsibilities of the engineer or designer go beyond simply maintaining professional standards and legal compliance (such as whistleblowing in the face of malpractice), but that questions of cultural, social and moral values should be formally *integrated* into design practice itself. Engineering ethicists have suggested that because technological innovations have embedded values within them (Urquhart et al., 2009) technology assessment should involve anticipatory technology ethics (ATE) (Brey, 2012) that involves foresight over emergent technological trends and their impacts, and ethical reflection embedded in the design process itself. Moreover, the converse is also true: 'designerly thinking' is a core component of ethical thinking (Dorst & Royakkers, 2006; Van Wynsberghe & Robbins, 2014). An integrative ATE necessarily entails practical knowledge, or knowhow (sometimes *techne* or 'craftsmanship') which complements, rather than is substituted by, an understanding of ethical theory, principles and rules. Conversely, for established technological artefacts, what is required is an ethics of *disclosure*—to uncover the embedded values within design processes and emergent socio-technical systems (Brey, 2000) such that governance can catch up with new and emergent socio-technical systems in which the ethics remain hidden from public scrutiny.

In both anticipatory and reflective scenarios, philosophers of technology have effectively challenged a purely instrumental vision of technological artefacts—whereby technologies are a means to an end, and thus the choice of technological means to solve problems is a morally neutral affair (Van De Poel, 2001) that can be left to market forces to control. As Rapp (1981) argues, purely instrumental visions of technology are inadequate because the formulation of the (moral) goals to be met by a technology cannot be separated from the development and choice of technological means to meet those goals. Moreover, technologies are not always developed with clear goals in mind and can influence the social and moral choices *after* they have been realised, released for broader consumption and then broadly disseminated within a technological culture (as per the Collingridge dilemma). It must be noted that such 'surprises' can be positive for society, for example aspirin and penicillin were both advancements that had profound changes to medical practice in the treatment of pain and infection respectively, though neither was specifically 'invented' with those goals in mind. It is important therefore to avoid conflating *unanticipated* and *undesirable* consequences of technological innovation (Healy, 2012). We can understand there is a core *metaethical* principle within the paradigm of technology governance, that discussion of ethical concerns should be a core aspect of the process of design. Yet within this, the technology remains largely neutral—the nature of the artefact is shaped by a deliberative process of ethical reflection which remains external to it. However, when it comes to technologies that facilitate direct human-computer interaction such as those involved in the creation of virtual realities, it behoves philosophers and technologists to think about ethics less as a process that shapes technological design, but rather one that is similarly interactive—a bidirectional process in which the technology plays an active role in shaping ethical reflection and the outcomes and consequences of human action.

The second implication is that the SCOT as a participatory-deliberative democratic process is one that necessitates the input of diverse value judgements, opinions and perspectives of multiple users, policymakers and affected citizens (the 'stakeholders'). Given the aforementioned governance challenges presented by SECTs, the need to *engage* with heterogeneous 'publics' in the early stages of technology development is now recognised, at the very least, as good practice (Felt & Fochler, 2008), and to others as a moral imperative (Fiorino, 1990) to ensure procedural justice (Joss & Brownlea, 1999)—fair access of affected parties to decisions that directly or indirectly affect their welfare. Direct democratic participation requires mechanisms to facilitate active involvement in decisions (Delgado et al., 2011; Charnley-Parry et al., 2017), design practices and SCOT implementation, not an abstract, arm's length consultation, which may simply be ignored by those in power (Smith, 1987; Wehling, 2012). The concept of technology assessment is now strongly associated with the participatory-deliberative turn discussed above. However, the strong democratic control gained through multi-stakeholder participatory-deliberative processes raises another problem for technology ethics. The notion that ethical reflection should be a component of technology governance is broadly established, though there remains little consensus amongst scholars as to the kind of ethics that should be practised, nor the individual(s) selected to perform this ethical analysis (Van Wynsberghe & Robbins, 2014). Simply bringing together a group of stakeholders and asking them to express, negotiate and apply their moral values (or perhaps asking them to rank judgements or choices) in relation to VR is not sufficient to ensure that ethical issues are satisfactorily resolved. There is a tension between what is *socially acceptable* and what is *ethically justified*. The two are related but are not the same thing.

The rise of the RRI platform and the governance of technology through participatory-deliberative mechanism of technology assessment raises David Hume's distinction between *description* and *prescription* in ethical analysis—the so-called Is-Ought conundrum from the Treatise on Human Nature (Hume, 1739). As Hume argued, understanding and categorising *moral systems, principles and values within groups, cultures and societies is not the same as judging their ethical validity. It is a mistake to simply ask what people think is right or wrong without either trying to improve the substantive quality of ethical judgements, or else evaluate the nature of the responses given* (Edel, 1998; Searle, 1964). Hume's Is-Ought conundrum has separated empirical (social) science from ethical analysis. In simple terms, we have come to understand the empirical as dealing with facts and ethics as dealing with norms and values. Empirical science is descriptive, ethics is prescriptive. In Moore's philosophy, to conflate facts and values is to commit a variation of the *naturalistic fallacy*: one cannot define the concept 'good' in terms of a natural property because it is a simple concept, that is one that cannot be defined in terms of any other concept, therefore one cannot derive what is ethical from other descriptive categories (Moore, 1903; Bruening, 1971). Conversely the *moralistic fallacy* is the assertion that moral judgements are of a different order from factual judgements: if one expressed a conclusion on what *is*, based only on what one believes ought to be (d'Arms & Jacobson, 2000; Moore, 1957). A publicly acceptable technology governance decision is usually defined as one whereby the majority agree on a course of action, however, this does not *automatically* imply adequate ethical justification. The popularity of a decision alone is an insufficient gauge of its ethical acceptability. *In situations where technology governance requires a combination of ethical assessment and pluralistic dialogue amongst diverse stakeholder opinions, what is needed are processes to balance the diverse moral judgements of an array of actors and to measure their acceptability amongst the group, but also to provide structure and ontological validity to those ethical judgements through processes to weigh evidence, facilitate ethical reflection, to choose amongst a range of options and hence reach ethically considered and consistent judgements about the problem at hand.*

Applied Ethics and Practical Ethics

The is-ought conundrum, naturalistic and moralistic fallacies have steered normative moral philosophers away from discussing individual subjective moral judgements, political institutions, cultural influences, and constraints, due to a fear that normative value will be lost to emotivism and moral relativism (Benedict, 1999). However, the problem remains that if ethics wants to say things about the real world, it has to take into account facts, subjective perspective and decision-making contexts if it is to be of use (Widdershoven & Van der Scheer, 2008; Alcorn, 2001; Haldane, 2012; London, 2001). As such, there is a growing interest in the idea of *integrating* empirical and normative research so that normative guidelines can be both established on the basis of empirical knowledge and evaluated through the analysis of the observable consequences of moral action (van der Scheer & Widdershoven, 2004). In essence, if ethics is to be inherently *practical* then an approach is required whereby ethical reflection and judgement happens in the context of our daily lives, the constraints and opportunities we face and the courses of action available to us.

Defence of this approach comes from Aristotle's concept of ethics as a practical science. Aristotle argued that ethics is the practice of virtues oriented towards the achievement of "the good life", something achieved as a function of individuals within society pursuing personal, intellectual and interpersonal excellence (Curzer, 2012). Individuals must use their judgement to evaluate the situations in which they were immersed, and thus take choices according to the will of being a good human being (Aristotle, 2000; Broadie, 1991). The good cannot be derived from empirical statements about what one believes to be good, but the action of *being good* does not take place in a vacuum, devoid of interpersonal relationships, economic considerations or material constraints. Understanding what it means to be good requires the contextualisation of moral principles, rules and theoretical considerations with an empirical understanding of the social features of human activity. Ethics must be applied in context. As such, the formal philosophical discipline of *applied ethics* is associated with implementation of moral principles, theories or rules to key areas of public and private lives in concert with empirical contextual analysis. Applied ethics is commonly differentiated from *normative ethics* (concerned with the development of moral rules and maxims for living an ethical life) and *metaethics* (concerned with the conditions under which moral decision-making occurs and the capacities of moral agents). It must be noted, however, that all of these elements are interrelated (Frey & Wellman, 2008; Winkler & Coombs, 1993) as discussed below.

Applied ethics emerged in part due to the social changes that occurred in the early 1970s due to rapid technological advancement (with respect to available medical treatments, changes in economic and business practices and the growing threat of global ecological catastrophe). Spurred by a sense of urgency driven by rapid change, applied ethics grew in popularity as a means to take philosophy 'out of the classroom' and into the broader public sphere where issues of practice, institutional and cultural constraints, and actor decision-making responsibility influence the context in which moral principles have relevance for ethical analysis. Applied ethics is thus ontologically different from a range of other 'applied' academic disciplines, such as applied mathematics, that apply theory to real-world phenomena but are not, in turn influenced by such phenomena. Many applied disciplines treat theory as something different to application. Such an applied ethics would treat normative moral philosophy—the theory and principles of ethical behaviour—a priori to action. The theorising takes place first and then the insight from the theory is put into practice. Such a top-down approach has fallen out of favour. MacIntyre and Beauchamp, for example, have questioned whether such a separation between theory and practice is useful for making ethical decisions at all. They argue that it is a mistake to think of ethics as a body of theory that can be wheeled in, when necessary, to sort out any particularly 'real-world' dilemma (MacIntyre, 1984a, 1984b; Beauchamp, 1984). As such, applied ethics is necessarily multidisciplinary and often related to the empirical insights gleaned from the social practices of those that work within (or are affected by) the social context in which specific domain-related ethical issues are raised.

The epistemological assumption at the heart of a practical ethics is that philosophical principles cannot always be applied in any straightforward way to particular social problems, technological developments or public policies. In the face of concrete dilemmas, one must revise philosophical principles in the face of social and technological change, as much as rely upon them for justification of specific courses of action. Within this framework of moral philosophy, reasoning is not the sole element of deliberation about practical questions. Other elements such as moral *intuition* and perception (the ability to recognise an ethical issue in a complex set of circumstances), moral imagination (to formulate and resolve the consequences of action) and moral *character* (the disposition to live ethically in a coherent way over time) are of equal importance (Thompson, 2007; LaFollette, 2003; Fesmire, 2001). Practical ethicists call for a more pragmatic view of the relationship between normative ethics (concerned with the justification of ethical action, the balancing of moral claims and the questions that arise when considering how one ought to act) and the practical decision-making that takes place in complex and uncertain situations, where the consequences of action are not immediately clear, there are multiple values and interests are at stake, and where changing situations call for a shift in our understanding and interpretation of ethical principles.

Ethical Tools

Practical ethicists have long been concerned with ways in which the clarity of moral reasoning derived from normative moral philosophy can be 'packaged' in such a way that non-philosophers can utilise the wisdom derived from formal ethics in practical situations that require such ethical insight. This type of practical ethics requires mechanisms for use by non-specialists. As such, the notion of ethical tools, toolkits, frameworks, instruments or methods (simplified here as 'ethical tools') have gained a degree of popularity in the literature on ethics education and practical ethics, particularly from the work of Weston (2000), Deblonde et al. (2007), Mepham (1999a) and Baggini and Fosl (2007). Tools are posited as support mechanisms to assist ethical judgements in situations where there is no direct recourse to normative moral theory, participants are untrained in ethical analysis, or the input of philosophers is unavailable or inappropriate. Tools are often necessary because ethical decision-making in professional contexts usually involves deliberation by non-philosophers. Such actors commonly recognise that a course of action has a moral component but may not know how to proceed. An ethical tool is designed to both simplify problems and solutions, and yet also provide sufficient scope for *critical thinking* and ethical reflection upon the problem at hand. Ethical tools can thus be thought of as a mixture of thought procedures, practical methods or "...judgment aids that help justify value choices without recourse to substantive theories or value systems of limited scope" (Forsberg, 2007, p. 456). Or as Moula and Sandin (2015) as:

> A practical method and/or conceptual framework with the main purpose of helping the user(s) improve their ethical deliberations in order to reach an ethically informed judgment or decision.

Ethical tools have two primary formats. The first are *judgement aids*, often taking the form of decision-trees, checklists or choice models. Such judgement aids are, broadly speaking, grounded within the *act-deontology* tradition of normative ethics (Vitell & Nin-Ho, 1997): centred upon the role of the individual actor to choose amongst available options through reflection upon personal judgements and the application of different (and oft-conflicting) principles, rather than adherence to a specific set of rules (rule-deontology) (Gaus, 2002; Spielthenner, 2005). At the heart of an ethical tool-based approach is a concern with *heteronomy*—that individuals make moral judgements without recourse to reason. In other words the judgements that they make are dictated by social norms and conventions rather than an internal deliberation upon different courses of action. The tool is used to improve the *autonomy* of the individual to make ethical decisions by providing a supportive framework to encourage reflection and evaluative judgement. Ethical decision-models tend to be grounded in the act-deontology tradition because they commonly concern the ethical implications of personal action in a professional setting (as a healthcare practitioner, business owner, engineer and so on); and the tool facilitates moral judgement by breaking it down into discrete steps. Many methods of ethical decision-making are designed to facilitate and structure this process of moral reflection in order to consider different inputs, perspectives and eventualities before coming to a decision on how to proceed. Given the breadth of these models across a range of applied ethics disciplines (primarily from medical, business and legal ethics) I will not consider each in turn, but rather draw out an emergent generic structure (for details of relevant models please refer to: ERC, 2004; Thomson, 1999; Marshall, 1999; Forester-Miler & Davis, 1996; Van-Hoose, 1980; Bowen, 2005; Potter, 1999; Jones, 1991; Park, 2012). Across these models, individual actors are encouraged to move sequentially through a series of evaluation stages in order to reach an ethically informed decision. Most of these models involve assessing relevant information followed by normative-theoretically informed reflection before reaching a decision. Some recognise that by completing and evaluating the ethical implications of action that this, in turn, raises new ethical questions for consideration. The general format of these checklist approaches is very broadly summarised in Fig. 3.1.

The models summarised in Fig. 3.1 are primarily for non-specialist actors in different professional situations trying to make decisions within the ethical norms of their respective organisation. The emphasis is upon individual judgement. Given that most professional actors in healthcare, business or legal settings (for example) are not professional ethicists, an ethical tool will usually employ some form of 'common sense principlism'—that is actors are presented with a selection of principles that are broadly understood and thus have a degree of support from a variety of

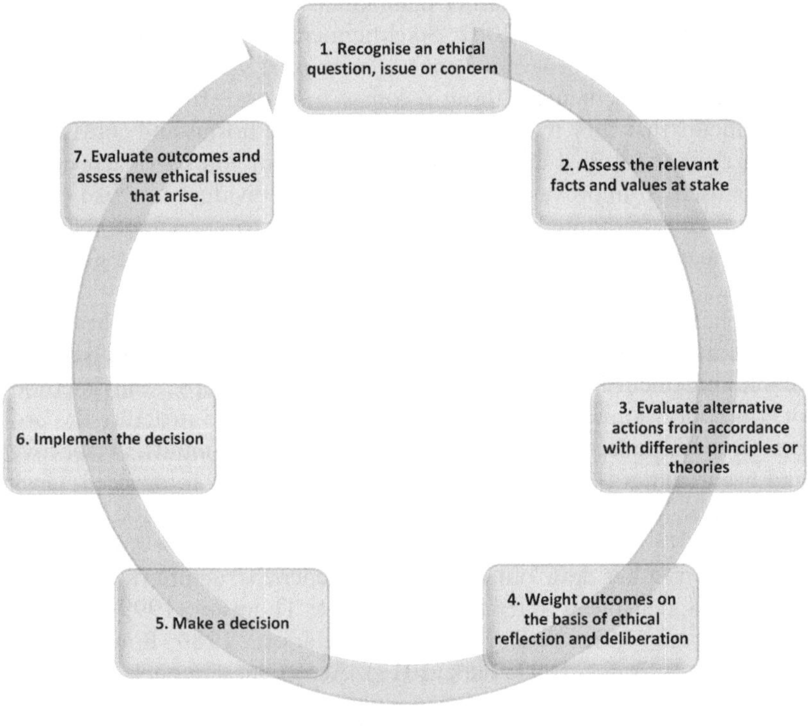

Fig. 3.1 Generic ethics decision-model

ethical theories and cultural beliefs (Beauchamp & Childress, 2001; Howard et al., 2002; Schmidt-Felzmann, 2003). *Principles* are useful distillations of normative theoretical positions. The aim of a principlist approach is commonly to improve the philosophical foundation of judgements made by non-specialist actors, given that "all evaluative judgments and more specifically all moral judgments are nothing but expressions of preference, expressions of attitude or feeling" (MacIntyre, 1984a, p. 12). As Kaler (1999) notes, this is a non-reductionist ontology in that it seeks to take into account a broad array of phenomena (including technical, scientific, principle and judgement-based factors) without reducing them to one or two core notions. The principlist approach to ethical toolmaking therefore attempts to do justice to a great variety of human experiences (ibid.). Notable in this regard is the work of Beauchamp and Childress. They distil complex ethical theories such as egalitarianism, deontology, utilitarianism, and virtue into simpler forms that are more easily understood and thus applied to real-world situations. Principlism is thus an extension of the Rawlsian tradition of using a 'common sense rule' (Schroeder & Palmer, 2003; Rawls, 1951) by applying prima facie, 'common sense' ethical principles. These include principles such as *autonomy* (respecting the decision-making capacities of persons), *non-maleficence* (avoiding the causation of harm), *beneficence* (providing net-benefits) and *justice* (distributing benefits, risks and costs fairly) (Beauchamp & Childress, 2001). Such a principlist approach has proved highly valuable to practise due to the simplicity, comprehensiveness and applicability of the approach to healthcare and bioethics settings.

Though common to individual decision-making in professional ethics settings, ethical tools can also take the form of *deliberative methods*—providing a mechanism to aid group decision-making, policy input and design processes for new technologies and plans through stakeholder dialogue on ethical content. Ethical tools such as those developed in the Ethical Bio-Technology Assessment 'tools' development process (Ethical Bio-TA project) (Kaiser et al., 2004) are mainly *participatory* methods used in ethical evaluation in different group-based policymaking contexts, primarily around issues of governance in agriculture and food production (Forsberg, 2007; Beekman & Brom, 2007; Kaiser et al., 2004, 2007; Kaiser & Forsberg, 2001). What Kaiser et al. (2004) argue is that procedures for analysing ethical issues must operate as structured decision-support frameworks suitable for assisting stakeholders in their deliberation, and then integrating the outcomes of the deliberation into design, implementation and policymaking. It is through the application in practical decision-support or policymaking that deliberative processes on ethics become 'tools' (Beekman & Brom, 2007). Tools such as the *ethical matrix* (Cotton, 2014a, 2009; Food Ethics Council, 2005; Mepham, 1999b; Kaiser & Forsberg, 2001; Schroeder & Palmer, 2003), the *ethical grid* (Seedhouse, 1998) and *reflective ethical mapping* (Cotton, 2014b) to assist stakeholders in making decisions by reflecting upon a range of ethical principles in relation to judgements, a range of affected stakeholder groups, and other contextual factors. In the aforementioned ethical tools for both individual and group decision-making, it is notable that none of them are based upon a single normative ethical theory, but rather encourage deliberation amongst a range of theoretically informed positions in concert with ethical reflection. Ethical tools are therefore useful in creating what Rawls terms *reflective equilibrium* (Daniels, 1979; Rawls, 1951)—a *coherentist* ethics whereby personal moral judgements and ethical principles are brought into harmony with one another through a process of reflection and deliberation about the relationship between personal values, ethical norms and broader institutional frameworks in which ethical judgement is applied. It is through deliberation, rather than the rigid application of specific normative rules that ethical judgement is refined.

Computer-Mediated Ethical Tools

In the following chapters VR is assessed as a potential platform for the development of a novel ethical tool-based approach. However, a number of other computerised systems for ethical decision-making have been developed in recent years and each have specific advantages. Most commonly, computerised systems follow pen-and-paper equivalents in terms of structure and function. For example, developing a checklist, decision-tree or matrix model is relatively simple using an online questionnaire format that branches questions dependent upon prior responses. It is also possible to input text, images or video content to act as a stimulus for ethical reflection. Numerous prototypes have been developed and evaluated for this purpose. One notable example is the EthicsGame™S approach, a bespoke system designed specifically for ethical reasoning within classroom situations for courses in business ethics. The system is designed as a *serious game* (Michael & Chen, 2005). Gaming is a novel approach to stimulate ethical reflection. A game is, as Juul argues, a rule-based system that combines variable and quantifiable outcomes, where different outcomes are assigned different values, the player exerts some form of effort in order to try and influence the outcome, the player feels a sense of attachment to the outcome, and the consequences of the activity are optional and (to some extent) negotiable (Juul, 2010). Rules, negotiation and outcomes make games and simulations potentially valuable as the basis for ethical tools. Games are highly popular in a range of educational contexts because they create rule-bound, social and competitive environments that provide a way for learners to employ practical knowledge in managing indeterminate, open-ended situations, and to develop practical skills in negotiation, argument construction, strategy formulation, and analysis of evidence. This is useful, as Lloyd and van de Poel (2008) argue, as the structuring of rules, behaviours, normative principles and their application to real-world situations can, through gaming, employ intuitive emotional responses as well as clearly thought out lines of reasoning (Kuhn, 1998).

The bespoke EthicsGame™ builds upon this approach. It is composed of a range of online programme modules: the Ethical Lens Inventory™, for encouraging students gain awareness of their implicit ethical preferences; a series of experiential case studies to help to develop an understanding of a range of ethical approaches, Ethics Exercises™ to reinforce ethical concepts, and Hot Topics Simulations™—interactive scenarios that challenge learners to make decisions from multiple ethical perspectives. Collectively, these tools aim to encourage the user to recognise when ethical situations require resolution, to assess the different options for action, identify solutions, and then communicate those to interested stakeholder parties—thus individuals can explore ethical dilemmas from multiple perspectives, learn to consider stakeholder impact while making an ethical decision and then articulate their own processes of decision-making to third parties (Litzky, 2012; EthicsGame, 2018; Baird, 2005).

Though online tools such as the EthicsGame™ are designed specifically for classroom use or professional training in ethical reflection, there is also a growing literature on the role of broader consumer recreational computer games and playing environments in structuring ethical reasoning, emotional responses and decision-making. Indeed as Prensky (2006) argues, children and young people are attracted to computer game playing precisely because they provide a learning environment through which they can experiment in risk-taking, strategy formulation, collaboration and antagonisation; all of which are relevant to complex ethical decision-making. Games commonly provide situations and embedded choice architecture to encourage players to choose amongst multiple outcomes, which in turn have positive or negative consequences for in-game characters. This is, in essence a form of *virtual ethics*: the consequences are not real, but they are understood, imagined and felt by the player. One popular category is the role-playing game: either as a stand-alone single-player 'adventure' (popular examples include Elder Scrolls™, Mass Effect™, Dragon Age™, or The Witcher™ series) or as a shared interactive experience: a massive multiplayer online role-playing game (MMORPG, e.g. Guild Wars™, StarCraft™ or World of Warcraft™). Such gaming platforms encourage the player to adopt an avatar within an imagined game world which has its own sets of social rules, dilemmas and constraints. Games in this vein present players with morally charged choices throughout: some such as the early Bioware's™ Mass Effect ™ games actually track the moral choices of the player as they progress through the game's narrative, using a sliding scale to determine the difference between what they term 'paragon and renegade' character traits which map to moral the virtues and vices within branching dialogue options and character choices within the game. Others such as the acclaimed Witcher, Wild Hunt™, adopt a different approach whereby moral choice is ambiguous and the consequences of specific in-game actions which at first appear virtuous may have negative consequences further down the line. Different games adopt different normative principles to structure the underlying moral decision framework (in these two examples showing the difference between virtue ethics and consequentialism). It is significant, however, that the story elements of such role-playing games explicitly highlight ethical choices, and that the immersive nature of the game-playing activity stems directly from this choice.

Game designers often pose ethical dilemmas as a means for the player to reflect upon their own judgements and core moral values. However, players do not always make morally good choices, and are rarely punished in the game for making morally bad choices. Gamers may act benevolently towards non-player characters within the game, or else gain enjoyment from subverting the moral rules of society by acting maliciously. It is difficult however to establish a link between the moral characteristics of the player and the actions taken within the gaming environment. Given the overarching context of a game players may subvert or circumvent their own moral reflections in order to either experience the consequence of virtually immoral behaviour (as a matter of personal interest or role play), or else optimise specific game pathways (such as to complete a game faster or receive some other in-game reward such as new items, or character experience that leads to other forms of game progression). It is therefore problematic to make assumptions about the moral characteristics of specific players based on their in-game actions.

Games provide two important elements for this analysis. First, is the capacity to create a world that has moral rules and social context. The virtual environments of gaming worlds are increasingly sophisticated, complex and interactive—though most take place in exotic environments (e.g. high fantasy settings, futuristic sci-fi scenarios or dystopian landscapes), others such as the Grand Theft Auto™ or Red Dead Redemption™ series track relatively close to contemporary or historical environments, whereby realism of social interaction within socioculturally and politically familiar environments is part of the appeal. Virtual environments found in gaming are shown to convincingly simulate the real-world responses of different actors to complex ethical situations. The narrative that they provide is laden with ethical choices, though these are of course implicit. Tools such as the EthicsGame™ are better structured for overt ethical deliberation but are less immersive: using textual representations such as emails to provide clarity and context to decisions that players make within the game. One must consider whether a combination of these two elements is possible—to combine the rich narrative appeal of commercial role-playing games with the philosophical clarity of a formal ethical tool. The use of VR would provide much greater realism, simulated interaction and immersion in a way that current formal ethics games do not. This leads to the second advantage of ethics-as-gaming. One of the core assumptions at the heart of this book is that emotional engagement with ethics is an important element, not only to improve the decision-making *quality* of the choices made, but also to stimulate a greater sense of personal attachment and investment in the outcomes of those decisions. If an individual learning about the ethical consequences of their professional activity can see first-hand in a simulated virtual environment what the impact that those choices have, then as with other training activities discussed in Chap. 2, this can enable embedded learning which carries over to real-world practice.

Mainstream role-playing games are not ethical tools in themselves, because the reflective process is not designed to 'stick' with the player. Game designers usually want you to be entertained rather than to become better people. Subversion of ethical norms is a valid playing style. What we can take away from these games, however, is the complexity of ethical decision-making that takes place within a naturalistic setting—players make choices within a world that has rules and conventions of its own. Adapting this approach for a VR-based ethical tool is possible. By directly relating the ethical decision-making that happens within the virtual environment to the moral attributes and decisions of the user made in real-world contexts, the complex choice architecture of a game-like environment can be utilised for practical ethics.

Conclusions

Virtual reality (VR) like any technological development presents new opportunities and challenges for established systems of ethical rules, norms and social conventions. The problem is that VR creates new forms of ethical heteronomy—the decisions that people make are dictated by new forms of virtual environment-specific social conventions and norms rather than deliberation and moral judgement. The challenges presented in Chap. 2 concerning sexist and racist behaviours, doxing, bullying or virtual assault require external scrutiny, moderation, rule-making and design adjustments to protect users. It is necessary therefore to open up the assessment of virtual environments to participatory processes of technology governance in order to avoid the negative social consequences of unregulated VR spaces.

Ethical heteronomy can also be alleviated by VR. In this chapter, different tools and frameworks to assist ethical judgement are discussed. The notion of an ethical tool is such that it alleviates the burden of structured deliberation on ethical matters by facilitating the process of reflection and application to real-world dilemmas. For those who are not themselves moral philosophers but are nonetheless experienced professionals that must make explicit ethical judgements (such as those in the fields of medicine, business or politics) ethical tools become vital heuristic devices to aid decision-making. Ethical tools can address the psychological skill aspects of ethical competence, to facilitate an awareness of ethical situations and conflicts, to understand competing values and actor perspectives, to support and sustain ethical deliberation, clearly express outcomes and consequences, and to coherently and transparently communicate the process and outcomes of ethical decisions. Computer-facilitated approaches can enhance these elements by simulating cases within virtual worlds, presenting a range of easily accessible alternative scenarios, facilitating group discussion and deliberation, providing access to a range of ideas, cases and theories, and reducing the workload and psychological burden associated with bringing these elements together (see, e.g. Kavathatzopoulos, 2003). Yet many of the existing decision-making tools available for computer-mediated ethical evaluation simply transfer pen-and-paper checklist and rule-based models into a digital environment. The richness of naturalistic decision-making, context and personal reflection is missed in questionnaire and matrix-based approaches. What VR allows is a stronger sense of immersion, facilitating new forms of reflection involving imaginative engagement with ethical dilemmas, interpersonal interaction, and crucially *empathy*. In Chap. 4, the rule-based and checklist approaches are critically evaluated with insight from feminist and pragmatist philosophical critique, and insights are drawn to the development of a novel VR-based ethical tool.

References

Alcorn, P. A. (2001). *Practical ethics for a technological world*. Prentice Hall.

Aristotle. (2000). *Nicomachean ethics* (R. Crisp, Trans.). Cambridge University Press.

Augustine, D. L. (2018). *Taking on technocracy: Nuclear power in Germany, 1945 to the present*. Berghahn Books.

Baggini, J., & Fosl, P. S. (2007). *The ethics toolkit: A compendium of ethical concepts and methods*. Wiley-Blackwell.

Baird, C. A. (2005). *Everyday ethics: Making hard choices in a complex world*. Tendril Press.

Beauchamp, T. L. (1984). On eliminating the distinction between applied ethics and ethical theory. *The Monist, 67*, 514–531.

Beauchamp, T. L., & Childress, J. F. (2001). *Principles of biomedical ethics* (5th ed.). Oxford University Press.

Beekman, V., & Brom, F. W. A. (2007). Ethical tools to support systematic public deliberations about the ethical aspects of agricultural biotechnologies. *Journal of Agricultural and Environmental Ethics, 20*(1), 3–12.

Beierle, T. C. (1999). Using social goals to evaluate public participation in environmental decisions. *Policy Studies Journal, 3*(4), 75–103.

Beierle, T. J. (2002). The quality of stakeholder-based decisions. *Risk Analysis, 22*(4), 739–748.

Benedict, R. (1999). A defense of ethical relativism. In H. J. Curzer (Ed.), *Ethical theory and moral problems*. Wadsworth Publishing Company.

Blowers, A., & Sundqvist, G. (2010). Radioactive waste management—Technocratic dominance in an age of participation. *Journal of Integrative Environmental Sciences, 7*(3), 149–155.

Bowen, S. A. (2005). A practical model for ethical decision making in issues management and public relations. *Journal of Public Relations Research, 17*(3), 191–216.

Brey, P. (2000). Disclosive computer ethics. *ACM Sigcas Computers and Society, 30*(4), 10–16.

Brey, P. A. (2012). Anticipating ethical issues in emerging IT. *Ethics and Information Technology, 14*(4), 305–317.

Broadie, S. (1991). *Ethics with Aristotle.* Oxford University Press.

Bruening, W. H. (1971). Moore and "is-ought". *Ethics, 81*(2), 143–149.

Charnley-Parry, I., Whitton, J., Rowe, G., Konrad, W., Meyer, J.-H., Cotton, M. D., Enander, A., Espluga, J., Medina, B., & Bergmans, A. (2017). *Principle for effective engagement. D5.1 for the history of nuclear energy and society project,* Brussels: European Commission.

Climate Assembly UK. (2020). *The path to net zero.* House of Commons with involve, Sortition Foundation and mySociety.

Collingridge, D. (1980). *The social control of technology.* Pinter.

Cotton, M. (2009). Evaluating the 'ethical matrix' as a radioactive waste management deliberative decision-support tool. *Environmental Values, 18*(2), 153–176.

Cotton, M. (2010). Discourse, upstream public engagement and the governance of human life extension research. *Poiesis & Praxis, 7*(1–2), 135–150.

Cotton, M. (2014a). Ethical matrix and agriculture. In P. B. Thompson & D. M. Kaplan (Eds.), *Encyclopedia of food and agricultural ethics* (pp. 1–10). Springer Netherlands.

Cotton, M. (2014b). *Ethics and technology assessment: A participatory approach.* Springer-Verlag.

Cotton, M. (2017). *Nuclear waste politics: An incrementalist perspective.* Routledge.

Curzer, H. J. (2012). *Aristotle and the virtues.* Oxford University Press.

d'Arms, J., & Jacobson, D. (2000). The moralistic fallacy: On the 'appropriateness' of emotions. *Philosophy and Phenomenological Research, 61*(1), 65–90.

Daniels, N. (1979). Wide reflective equilibrium and theory acceptance in ethics. *Journal of Philosophy, 76*(5), 256–282.

Deblonde, M., De Graafe, R., & Brom, F. (2007). An ethical toolkit for food companies: Reflections on its use. *Journal of Agricultural and Environmental Ethics, 20,* 99–118.

Delgado, A., Kjølberg, K. L., & Wickson, F. (2011). Public engagement coming of age: From theory to practice in sts encounters with nanotechnology. *Public Understanding of Science, 20*(6), 826–845.

Devon, R. (2004). Towards a social ethics of technology: A research prospect. *Techné: Research in Philosophy and Technology, 8*(1), 99–115.

Dorst, K., & Royakkers, L. (2006). The design analogy: A model for moral problem solving. *Design Studies, 27*(6), 633–656.

Edel, A. (1998). *Science and the structure of ethics.* Transaction Publishers.

ERC. (2004). *PLUS—A process for ethical decision making*. Washington: Ethics Resource Centre. Retrieved December 1, 2004, from http://www.ethics.org/plus_model.htm

EthicsGame. (2018). *EthicsGame: How it works*. EthicsGame. https://www.ethicsgame.com/exec/site/How_it_works.html

European Commission. (2018). *Responsible research and innovation*. https://ec.europa.eu/programmes/horizon2020/en/h2020-section/responsible-research-innovation

Felt, U., & Fochler, M. (2008). The bottom-up meanings of the concept of public participation in science and technology. *Science and Public Policy, 35*(7), 489–499.

Fesmire, S. (2001). Imagination in pragmatist ethics. In Proceedings of the Conference of the Society for the Advancement of American Philosophy, Vol. 11, p.13.

Fiorino, D. (1990). Citizen participation and environmental risk: A survey of institutional mechanisms. *Science, Technology & Human Values, 15*(2), 226–243.

Food Ethics Council. (2005). *Ethical matrix: Uses*. Food Ethics Council. Retrieved August 2, 2007, from http://www.foodethicscouncil.org/ourwork/tools/ethicalmatrix/uses

Forester-Miler, H., & Davis, T. (1996). *A practitioner's guide to ethical decision making*. American Counselling Association.

Forsberg, E. M. (2007). Pluralism, the ethical matrix, and coming to conclusions. *Journal of Agricultural and Environmental Ethics, 20*(4), 455–468.

Frey, R. G., & Wellman, C. H. (2008). *A companion to applied ethics*. John Wiley & Sons.

Gaus, G. F. (2002). What is deontology? Part One: Orthodox views. *Journal of Value Inquiry, 35*, 27–42.

Haldane, J. (2012). *Practical philosophy: Ethics, society and culture*. Imprint Academic.

Healy, T. (2012). The unanticipated consequences of technology. *Nanotechnology: Ethical and social Implications*, pp. 155–173.

Horlick-Jones, T., Walls, J., Rowe, G., Pidgeon, N., Poortinga, W., & O'Riordan, T. (2006). On evaluating the GM Nation? Public debate about the commercialisation of transgenic crops in Britain. *New Genetics and Society, 25*(3), 265–288.

Howard, B. J., Forsberg, E. M., Kaiser, M., & Oughton, D. (2002). 'An Ethical Dimension to Sustainable Restoration and Long-Term Management of Contaminated Areas'. International Conference On Radioactivity in The Environment, Monaco, 506–510.

Hume, D. (1739). *Treatise on human nature: Of virtue and vice in general*. Oxford University Press.

Jones, T. M. (1991). Ethical decision making by individuals in organizations: An issue-contingent model. *The Academy of Management Review, 16*(2), 366–395.

Joss, S., & Brownlea, A. (1999). Considering the concept of procedural justice for public policy-and decision-making in science and technology. *Science and Public Policy, 26*(5), 321–330.

Juul, J. (2010). The game, the player, the world: Looking for a heart of gameness. *PLURAIS-Revista Multidisciplinar, 1*(2), 248–270.

Kaiser, M., & Forsberg, E. M. (2001). Assessing fisheries—Using an ethical matrix in participatory processes. *Journal of Agricultural and Environmental Ethics, 14*, 191–200.

Kaiser, M., Millar, K., Forsberg, E.-M., Baune, O., Mepham, B., Thorstensen, E., & Tomkins, S. (2004). Decision-making frameworks. In V. Beekman (Ed.), *Evaluation of ethical bio-technology assessment tools for agriculture and food production: Interim report ethical bio-ta tools.* Agricultural Economics Research Institute.

Kaiser, M., Millar, K., Forsberg, E. M., Thorstensen, E., & Tomkins, S. (2007). Developing the ethical matrix as a decision support framework: GM fish as a case study. *Journal of Agricultural and Environmental Ethics, 20*(1), 53–63.

Kaler, J. (1999). What's the good of ethical theory? *Business Ethics: A European Review, 8*(4), 206–213.

Kavathatzopoulos, I. (2003). The use of information and communication technology in the training for ethical competence in business. *Journal of Business Ethics, 48*(1), 43–51.

Keulartz, J., Shermer, M., Korthals, M., & Swierstra, T. (2004). Ethics in technological culture: A programmatic proposal for a pragmatist approach. *Science, Technology & Human Values, 29*(1), 3–29.

Kleinman, D. L. (Ed.). (2000). *Science, technology and democracy.* State University of New York Press.

Konrad, W., Espluga, J., Bergmans, A., Charnley-Parry, I., Cotton, M. D., Enander, A., Meyer, J.-H., Rowe, G., & Whitton, J. (2018). *Comparative cross-country analysis on preliminary identification of key factors underlying public perception and societal engagement with nuclear developments in different national contexts.* Brussels: European CommissionD4.2 (2018 update).

Kuhn, J. W. (1998). Emotion as well as reason: Getting students beyond 'interpersonal accountability'. *Journal of Business Ethics, 17*(3), 295–308.

LaFollette, H. (2003). Pragmatic ethics. In H. LaFollette (Ed.), *Ethical theory.* Blackwell Publishing.

Law, A., & Mooney, G. (2012). Competitive nationalism: State, class, and the forms of capital in devolved Scotland. *Environment and Planning C: Government and Policy, 30*(1), 62–77.

Lidskog, R. (2008). Scientised citizens and democratised science. Re-assessing the expert-lay divide. *Journal of Risk Research, 11*(1–2), 69–86.

Litzky, B. E. (2012). Review of EthicsGame simulation. *Journal of Business Ethics Education, 9*, 485–488.

Lloyd, P., & van de Poel, I. (2008). Designing games to teach ethics. *Science and Engineering Ethics, 14*(3), 433–447.

London, A. J. (2001). The independence of practical ethics. *Theoretical Medicine and Bioethics, 22*(2), 87–105.

MacIntyre, A. (1984a). *After virtue: A study in moral theory.* University of Notre Dame Press.

MacIntyre, A. (1984b). Does applied ethics rest on a mistake? *The Monist, 67*, 489–513.

Maranta, A., Guggenheim, M., Gisler, P., & Pohl, C. (2003). The reality of experts and the imagined lay person. *Acta Sociologica, 46*(2), 150–165.

Marshall, J. (1999). *An ethical decision-making model: Five steps of principled reasoning.* Josephson Institute of Ethics. http://www.ethicsscoreboard.com/rb_5step.html

Mepham, B. (1999a). A framework for the ethical analysis of novel foods: The ethical matrix. *Journal of Agricultural and Environmental Ethics, 12*, 165–176.

Mepham, B. (1999b). A framework for the ethical analysis of novel foods: The ethical matrix. *Journal of Agricultural and Environmental Ethics, 12*(2), 165–176.

Michael, D., & Chen, S. (2005). *Serious games: Games that educate, train, and inform.* Muska & Lipman.

Mohr, A. (2007). Against the stream: Moving public engagement on nanotechnologies upstream. In R. Flynn & P. Bellamy (Eds.), *Risk and the acceptance of new technologies.* Palgrave Macmillan.

Moore, G. E. (1903). *Principia Ethica.* Cambridge University Press.

Moore, E. C. (1957). The moralistic fallacy. *The Journal of Philosophy, 54*(2), 29–42.

Moula, P., & Sandin, P. (2015). Evaluating ethical tools. *Metaphilosophy, 46*(2), 263–279.

Murphy, J., Levidow, L., & Carr, S. (2006). Regulatory standards for environmental risks: Understanding the US–European Union conflict over genetically modified crops. *Social Studies of Science, 36*(1), 133–160.

Nemet, G. F. (2009). Demand-pull, technology-push, and government-led incentives for non-incremental technical change. *Research policy, 38*(5), 700–709.

Nowotny, H. (2003). Democratising expertise and socially robust knowledge. *Science and Public Policy, 30*(3), 151–156.

Owen, R., Macnaghten, P., & Stilgoe, J. (2012). Responsible research and innovation: From science in society to science for society, with society. *Science and Public Policy, 39*(6), 751–760.

Park, E.-J. (2012). An integrated ethical decision-making model for nurses. *Nursing Ethics, 19*(1), 139–159.

Pidgeon, N., & Rogers-Hayden, T. (2007). Opening up nanotechnology dialogue with the publics: Risk communication or 'upstream engagement'? *Health, Risk & Society, 9*(2), 191–210.

Potter, R. B. (1999). The origins and applications of "Potter Boxes." Paper presented to the State of the World Forum, October 1999, San Francisco, CA.
Prensky, M. (2006). *Don't bother me, Mom, I'm learning!: How computer and video games are preparing your kids for 21st century success and how you can help!*: Paragon House St. Paul.
Rapp, F. (1981). The Technological World. In Analytical Philosophy of Technology (Springer, Dordrecht): 118–186.
Rawls, J. (1951). Outline of a decision procedure for ethics. *The Philosophical Review, 60*(2), 177–197.
Rip, A., & Kemp, R. (1998). Technological change. In S. Rayner & E. Malone (Eds.), *Human choices and climate change*. Battelle.
Rip, A., Schot, J. W., & Misa, T. J. (1995). Constructive technology assessment: A new paradigm for managing technology in society. In A. Rip, J. W. Schot, & T. J. Misa (Eds.), *Managing technology in society. The approach of constructive technology assessment* (pp. 1–12). Pinter Publishers.
Schmidt-Felzmann, H. (2003). Pragmatic principles—Methodological pragmatism in the principle-based approach to bioethics. *Journal of Medicine and Philosophy, 28*(5–6), 581–596.
Schroeder, D., & Palmer, C. (2003). Technology assessment and the 'ethical matrix'. *Poiesis & Praxis, 1*(4), 295–307.
Sclove, R. (1995). *Democracy and technology*. Guilford Publications.
Searle, J. R. (1964). How to derive 'ought' from 'is'. *The Philosophical Review, 73*, 43–48.
Seedhouse, D. (1998). *Ethics: The heart of health care*. Wiley.
Sheetz, T., Vidal, J., Pearson, T. D., & Lozano, K. (2005). Nanotechnology: Awareness and societal concerns. *Technology in Society, 27*(3), 329–345.
Smith, L. G. (1987). The evolution of public participation in Canada: Implications for participatory practice. *British Journal of Canadian Studies, 2*(2), 213–235.
Spielthenner, G. (2005). Consequentialism or deontology? *Philosophia, 33*(1), 217–235.
Stirling, A., O'Donovan, C., & Ayre, B. (2018). Which way? Who says? Why? Questions on the multiple directions of social progress. *Technology's Stories*, pp. 1–20.
Taebi, B., Correlje, A., Cuppen, E., Dignum, M., & Pesch, U. (2014). Responsible innovation as an endorsement of public values: The need for interdisciplinary research. *Journal of Responsible Innovation, 1*(1), 118–124.
Thompson, D. F. (2007). *What is practical ethics?* Harvard University.
Thomson, A. (1999). *Critical reasoning in ethics: A practical introduction*. Routledge.
Urquhart, C., Underhill-Sem, Y., Pace, T., Houssian, A., & McArthur, V. (2009). Are socially exclusive values embedded in the avatar creation interfaces of

MMORPGs? *Journal of Information, Communication and Ethics in Society, 7*(2/3), 192–210.
van der Scheer, L., & Widdershoven, G. (2004). Integrated empirical ethics: Loss of normativity? *Medicine, Health Care and Philosophy, 7*(1), 71–79.
Van De Poel, I. (2001). Ethics, technology assessment and industry. *TA-Datenbank-Nachrichten, 2*(10), 51–61.
Van Den Ende, J., Mulder, K., Knot, M., Moors, E., & Vergragt, P. (1998). Traditional and modern technology assessment: Toward a toolkit. *Technological Forecasting and Social Change, 58*(1), 5–21.
Van Wynsberghe, A., & Robbins, S. (2014). Ethicist as Designer: A pragmatic approach to ethics in the lab. *Science and Engineering Ethics, 20*(4), 947–961.
Van-Hoose, W. H. (1980). Ethics and counseling. *Counseling & Human Development, 13*(1), 1–12.
Vitell, S. J., & Nin-Ho, F. (1997). Ethical decision making in marketing: A synthesis and evaluation of scales measuring the various components of decision making in ethical situations. *Journal of Business Ethics, 16*(7), 699–717.
Wehling, P. (2012). From invited to uninvited participation (and back?): Rethinking civil society engagement in technology assessment and development. *Poiesis & Praxis, 9*(1–2), 43–60.
Wenk, E. (1975). Technology assessment in public policy: A new instrument for social management of technology. *Proceedings of the IEEE, 63*(3), 371–379.
Weston, A. (2000). *A 21st century ethical toolbox*. Oxford University Press.
Whitton, J., Brasier, K., Parry, I., & Cotton, M. (2017). The development of shale gas governance in the United Kingdom and United States: Opportunities for public participation and implications for social justice. *Energy Research & Social Science, 26*, 11–22.
Widdershoven, G., & Van der Scheer, L. (2008). Theory and methodology of empirical ethics: A pragmatic hermeneutic perspective. In G. Widdershoven, T. Hope, J. McMillan, & L. Van der Scheer (Eds.), *Empirical ethics in psychiatry* (pp. 23–36). Oxford University Press.
Wilsdon, J., & Willis, R. (2004). *See-through science: Why public engagement needs to move upstream*. Demos.
Winkler, E. R., & Coombs, J. R. (1993). *Applied ethics: A reader*. Blackwell.
Wynne, B. (1996). May the sheep safely graze? A reflexive view of the expert-lay knowledge divide. In S. Lash, B. Szerszynski, & B. Wynne (Eds.), *Risk, environment and modernity*. Sage Publications.

CHAPTER 4

Empathy and Ethics

Abstract In this chapter, the role of empathy and moral imagination is discussed in relation to the development of ethical tools. Empathy as a psychological and physiological process is discussed. It is then critically assessed as the foundational basis to moral judgement with reference to sentimentalist and feminist philosophies and critiques from moral psychology. The second part of the analysis outlines a model combining empathy and moral imagination through reflective judgement, arguing that this provides a sound basis for an ethical decision-making tool. The conceptual framework for this tool builds upon the work of pragmatist philosopher John Dewey through the model of what he termed ***dramatic rehearsal.***

Keywords Empathy • John Dewey • Pragmatism • Dramatic rehearsal

Introduction: Empathy and Ethics

Across the range of ethical tools discussed in Chap. 3, the emphasis lies with rule-based ethics—that ethical obligations originate in rational deliberation, some form of mutual agreement or contract between people to uphold the outcomes of ethical deliberation, and to establish reciprocity and mutual moral value. The wielder of an ethical tool is necessarily construed as a rational actor, one that inhabits an imagined world often

M. Cotton, *Virtual Reality, Empathy and Ethics*,
https://doi.org/10.1007/978-3-030-72907-3_4

projected by contractarian theories—as McCracken and Shaw (1995) argue—whereby humans are characterised by preference or value maximisation, and that ethical practice operates within a framework of clearly defined rules, rights and obligations. Yet as discussed in Chap. 2, and summarised by Pellizzoni (2012), new and emerging techno-sciences produce peculiar ontologies and messy ethical problems—the rational human agent is confronted with a contingent and indeterminate biophysical world that expands the scope of rational action. As a work of practical ethics, the conceptual framework for this book is situated within a debate on the relationship between empathy, imagination and messy real-world ethical problems. As discussed in the previous chapters, a well-designed VR provides an opportunity for the individual to become immersed in an environment, its characters and objects in a 'safe' way—one in which interactions can be controllable, repeatable, and potentially limited in their negative impact upon others. By providing the opportunity for an individual to encounter new and unfamiliar social interactions and moral choices, I argue that VR provides a unique environment through which to explore different novel ethical situations and new forms of reflection within a controlled environment. One facet of VR that is specifically of philosophical interest is the technology's capacity to improve the emotional attachment that users feel towards the decisions that they make. The key to this attachment is the capacity of the technology to inculcate feelings of *empathy* towards others through simulated interaction.

The human faculty of empathy is of philosophical importance because it allows an individual to have specific insight into the life of another. The psychologist and neurobiologist Baron-Cohen (2012) argues that empathy has two primary components: a *cognitive* part—which is how we understand other people, and an *affective* part that relates our emotional reactions to others. As Haney (1994) argues, empathy allows insight in a way that one cannot achieve through direct observation or evidence alone. For example, one cannot directly perceive the pain of another but can nonetheless empathise with the experience of pain. This is done either by relating our own experience to those reported by others; or else imagining ourselves in the other person's situation. We then reflect upon our own anticipated feelings as if we were to be put in the same situation. Empathy is therefore the naturally occurring subjective experience of similarity between one's own feelings and those expressed by others, whilst simultaneously recognising whose feelings belong to whom.

Neuroscientific analysis has shown that individuals will automatically share the emotions of others when they are exposed to these external emotional states (De Vignemont & Singer, 2006). Empathy involves not only the affective experience of the other person's actual or inferred emotional state but also some minimal recognition and understanding of another's emotional state (Decety & Jackson, 2004). Feeling empathy is an evolutionary sociobiological trait for humans (Zahn-Waxler et al., 1985; Decety & Jackson, 2004). Sharing the emotions of others has survival benefits when it leads to the sharing of resources, shelter and protection. Empathy stimulates altruistic behaviour by encouraging us to think beyond our own individual needs, or those of our immediate family, and to reorient our actions towards the benefit of the social group as a whole.

Hoffman (2000) discusses the relationship between children's cognitive development and the emergence of evermore sophisticated levels of empathic response, distinguishing empathy across a series of developmental stages, ultimately constituting a continuum of responses. The most basic is the immediate reactive response (such as the response to a newborn baby's cry), followed by egocentric empathic distress (e.g. being upset at the sight of someone else's distress and requiring personal comfort). At later stages of development individuals show more sophisticated responses, including empathy to others by providing comfort during their distress, and then empathy towards others beyond the immediate situation (such as recognising the distress of others following a natural disaster in another part of the world), and eventually empathy towards whole groups of people, including those identified as 'other' or 'out-group' individuals. It is clear that empathy varies in complexity and sophistication between individuals, and importantly these personal psychological qualities are broadly influenced by sociocultural, material and environmental factors—strong personal empathy is associated with interpersonal unity, such that the conception of self and other are in some sense merged together (Cialdini et al., 1997). To sustain this level of empathic response requires a supportive culture that favours altruistic behaviours across both in-group and out-group social interactions. Judging the empathy of any individual, therefore, makes sense only in the context of a broader set of social relationships, cultural norms and political values that foster altruism.

The relationship between empathy, the imagination of others' experiences and the decision-making of the individual is an important one within moral philosophy. The concept of *sympathy* was discussed in relation to ethics at least since the Scottish Enlightenment. Hume conceived of sympathy as the principle that allows one's moral judgements to emerge because it forms the source of one's moral distinctions. The concept of *moral sentimentalism* associated with the work of Hume (Hume, 1739; Collier, 2010; Ferreira, 1994) and Hutcheson (Hutcheson, 1725; Chismar, 1988), attempts to account for the meaning of moral phenomena by understanding the role that emotion and desire play in how individuals make sense of the practical aspects of ethical dilemmas. However, explicit reference to 'empathy' as a philosophical phenomenon emerged in the work of Theodore Lipps—who took the term 'Einfühlung' important to German aesthetics in the late nineteenth century and transformed it into a core component of social philosophy. Lipps (1903) argued from a phenomenological perspective that our perceptual encounters with aesthetic objects trigger internal psychological processes that give rise to experiences similar to those of physical movement or other forms of bodily sensation. One experiences a sense of vitality through an empathic connection with aesthetic perception. Lipps (ibid.) then asserts that empathy should be understood as the primary epistemic means to understand the minds of others—that empathy allows the possibility to know 'the other' and to provide insight into the lives of others without direct observation or experience. This idea was influential on the writing of Husserl (1931) and Stein (1917), who argued that empathy acts as an irreducible "type of experiential *act sui generis*"—it involves the prior recognition of separateness—that individuals recognise one another as discrete entities that are nonetheless 'knowable' through a conscious process in which one makes sense of others in their self-experience. In practical terms, empathy is the process through which an individual discloses their affective states, intentions and motivations and thus overcomes the separation between subjects.

In order to avoid moral heteronomy, whereby judgements are made simply through recourse to prevailing attitudes or norms of social behaviour, moral philosophers (particularly those within the Kantian tradition) require ethical decision-makers to be motivated by reflection, rational argumentation, impartiality and a rule-based approach, in order to provide ontological consistency. Living morally is thus principally a matter of moral insight into the application of moral rules, which is combined with the strength of moral character required to implement the wisdom of these rules to everyday life (Baggini & Fosl, 2007; Sumner, 1967; Garner & Rosen, 1967; Brandt, 1959; Lekan, 2006). However, a rule-based moral philosophy precludes a role for imagination and creativity in ethical reasoning—a fact that many moral psychologists have critiqued given that these characteristics are intrinsic to all other forms of human thought and action (Haidt, 2001; Haidt, 2003). Human motivation and moral reasoning as described through the realist accounts of human psychology find decision-making processes to be partial, partisan, limited in scope and swayed by numerous heuristics, biases and social practices in contrast to the normative ideals expressed in formal philosophical analysis (Dunn, 2004; Kohlberg, 1984; Haidt, 2001). Though Kantian philosophers eminently favour reason over sentiment and remain sceptical about empathy as a core feature of ethical decision-making, these features cannot be ignored because they are core elements of the judgements that we make day to day. Individuals believe things to be right or wrong and often this is a feeling that emerges fully formed within the imagination rather than a logical conclusion drawn from a rational deliberative process.

The notion of metaphysical sentimentalism in which moral facts make reference to one's emotive responses is relevant to the discussion of empathy. Sentimentalism is grounded in psychology, and so has empirical as well as philosophical antecedents. This goes back to Hutcheson (1725) who argued that a moral sense is innate, moral approval (a sense that something is inherently right or wrong) is seen in the sentiments of even very young children. Empirical research on moral behaviour has prompted renewed interest in the concepts of moral reasoning, of which empathy is an important component. Moreover, empathy is becoming part of mainstream public political discourse. Former President of the United States Barack Obama discussed the idea of an *empathy deficit* creating divisions within communities and across the political spectrum (Schumann et al., 2014), an issue brought into sharp relief during subsequent electoral processes in the United States. Empathy is ethically significant because it joins together psychological, philosophical and political dimensions. Hoffman (2000) describes empathy as a form of moral motive—it is a specific affective response which encourages the individual to act in another's interests, even at the expense of one's own. Empathy allows us to experience difference, thus making us question core aspects of our personal identity, values and beliefs. On the surface, empathy allows us to develop compassionate relationships with other people. On a deeper psychological level, it can also change the personal sense of who we are, and this stimulates corresponding changes to the duties we hold towards others and the types of social and political values that we project into the world (Clohesy, 2013); namely it can motivate prosocial behaviour towards a more general ethics of care.

Feminist Ethics and Empathy

Empathy in public life is articulated within feminist philosophy. Certain feminist philosophers emphasise the role that empathy and care play in the construction of moral rules and principles. Notably, Nussbaum (2001) states that it is men who like to invent elaborate abstract formal systems of moral thought which are then imposed upon the somewhat messier world of human moral relationships. A feminist conceptualisation of ethics highlights how the male-philosopher dominated contractarian and rationalist paradigms of ethics have tended, in the first instance, to prescribe particular attributes to moral actors: a set of characteristics deemed necessary for the practice of ethics. At the top of this list of attributes is deductive reasoning. Traditional normative ethics is centred upon moral actors displaying characteristics associated with a set of 'masculine' attributes such as courage, endurance, rational thought, 'wiliness' and sound political and economic judgement. An emphasis upon these traits subjugates women's ethical judgements as being based upon timidity, tenderness, compliance, docility, innocence, nurturing and so on (Almond, 1988). Women's characteristic contributions to ethical reasoning have, therefore, long been positioned as being of peripheral importance. This is due to an arbitrary dualism between the so-called objective, public-focused, rational and contractual aspects of a male-associated tradition of ethics when positioned against empathic, nurturing and social relationship-focused ethics associated with women. As Gilligan argues: "men tend to embrace an ethic of rights using quasi-legal terminology and impartial principles ... women tend to affirm an ethic of care that centres on responsiveness in an interconnected network of needs, care, and prevention of harm. Taking care of others is the core notion" (Gilligan, 1982 cited in Beauchamp & Childress, 2001, p. 371). Such care is only possible when one can empathise with the experiences of those that require such care (Slote, 2007). Slote (2007, 2017) further argues that empathy and care are consonant with Aristotelian virtues—they are preconditions for moral behaviour and the fair distribution of resources. A feminist care ethics is therefore tightly bound in the moral psychology of empathy.

Feminist ethics is a diverse field, though at the risk of oversimplification, there is a common critique amongst feminist philosophers of systems of ethics built upon the establishment of social contracts, the clarification of personal maxims, moral guidelines and hypothetical moral dilemmas, when these override the specificities of actual social and moral reality

(Sinnott-Armstrong, 1987; Shrage, 1994; Brennan, 1999). Collectively, feminist ethics challenges the systemic "depersonalising of the moral and demoralising of the person" (Walker, 1989, p. 16). Feminist insight to ethical practice is relevant to the ethics of VR as a reflective tool, because it reveals how personal experience, relationships and context are key aspects of ethical decisions (Porter, 1999). A VR-based ethical tool must therefore facilitate a process of relationship-building. Understanding the individual, their circumstances and goals is important in illustrating why different individuals would make different moral choices under the same circumstances. Within the principlist approach to ethical tool-making discussed in the previous chapter, the principles act as a heuristic device drawing upon the thinking of well-established theoretical traditions in order to decipher moral courses of action. But to feminist philosophers such abstract principles are an insufficient base to sound moral action. An understanding of the concrete specifics of actual ethical judgements, cases and the actors within decisions is vital, and so a broader understanding of context is essential (Doppelt, 2002; Brennan, 2002).

Critiquing Empathy

Empathy implies both the cognitive capacity and will to understand the perspectives of others, what it means to experience their thoughts and desires, to share their emotions, and care for them. As Baron-Cohen (2012) argues, empathy becomes the ultimate solvent, in that all problems can at least be helped by one taking a more empathic stance towards another. There is a tendency to presume that empathic reasoning is in some sense a moral virtue, though Battaly (2011) notes that the ordinary concept of empathy is poorly theorised as a moral philosophy.

Bloom (2017a, 2017b) argues that empathy has a positive role for society because it forms the basis of human solidarity. However, it is weaker as a basis for moral decision-making. He argues that empathy is positive but only under certain restricted circumstances, it acts as a spotlight, favouring interpersonal relationships between the ethical decision-maker and another individual, which is fine if there are only two people involved in the ethical dilemma. However, using empathy to guide our reasoning means we tend to care more about a single person than a whole group, perhaps even a whole population. Ethical questions of utility, just resource distribution, fair treatment, due process, duty and reciprocity cannot be adequately explained through the capacity of one to feel the emotions and feelings of another, and then act morally as a result. In other words, empathy as the basis of moral reasoning leads us to prioritise the needs of the few over the many, which can lead to ethical mistakes. Bloom also argues that many interpersonal relationships require the putting aside of feelings and empathy in order to make hard choices in the best interests of the moral subject. The actions of healthcare professionals in emergency situations, for example, require a certain amount of emotional distance in order for the practitioner to do what is best; shutting down their empathic response to the situation in order to do what is morally good.

From a psychological perspective, making moral decisions on the basis of empathy is problematic due to the interpersonal biases that are grounded in our ambient culture, social practices and political values. Humans have a tendency to share empathic relationships mediated through frames of similarity and proximity: with those that are *closer* to us rather than those that are further away. This similarity and proximity might be physical, geographical or emotional (such as to our family, friends and neighbours); however, it might also be based upon arbitrary features such as physical appearance, language or identity. Individuals tend to adopt preferential attitudes towards those that they identify as being similar to themselves (the in-group) often at the expense of those that they identify as different (the out-group). Being empathic towards the in-group may therefore come at the risk of dehumanising (or at least deprioritising) the out-group. Songhorian (2019) argues, therefore, that we should view empathy as an amoral ability from which one can develop either moral or immoral character—because empathy itself can be easily biased by factors that are not morally relevant such as ethnicity, gender, sexual orientation, nationality, and so on. Individuals will therefore have unjustified preferences towards those that are socioculturally and geographically proximate and similar, which is not ethically justified. Empathy alone is insufficient to act and judge morally and thus other moral qualities, such as duty, fairness, justice and impartiality are needed to modulate one's empathic response. As such, Bloom (2017a) argues that we should replace empathy with *rational compassion*—that our judgements should be guided not just by shared emotional states, but by a desire to care for others based upon reflective judgement over one another's needs.

Empathy can also be pathological—sadism, for example, is an empathic response to the distress of others that is overridden for personal pleasure at the expense of another. Similar experiences such as *schadenfreude* share a similar psychological root. The psychopathology of psychopathy is a case in point. Individuals diagnosed with psychopathy will often have few or no observable empathic traits, but this does mean that they will *a priori* conduct immoral behaviours. One without empathy can still obey ethical rules, and thus empathy is not a precondition for moral behaviour. Similar questions regarding individual moral agency arise for people with autistic spectrum disorder (ASD) who may not display behaviours that are traditionally associated with empathy (Songhorian, 2019) but this does not impact the morality of their choices. Thus empathy alone is inadequate for describing the moral character of individuals and therefore has a contested ontological status in moral philosophy. Empathy has been defined either as a precondition for moral behaviour, as a virtue Slote (2007), or else as an amoral psychological characteristic that requires guiding principles to ensure good behaviour (Songhorian, 2019). We can conclude therefore that additional cognitive and affective faculties are needed to moderate and guide empathic responses in order to lead to morally positive outcomes. Moreover, in the context of this analysis, we must question how can these faculties be supported within a virtual environment?

Empathy and Moral Imagination

It is clear from the analysis of Bloom (2017b) that empathy alone as a neurobiological and social construct is insufficient to ensure morally justifiable outcomes; empathy is better understood as a psychological trait rather than moral virtue. An ethical tool grounded in empathy-arousal therefore requires a second mechanism to correct the moral mistakes that may emerge due to cultural bias, the emphasis upon individuals, proximity and similarity. This second mechanism is ***moral imagination***.

The concept of moral imagination is traceable to the work of David Hume (Ferreira, 1994; Heath, 1995). Hume began with a concern for the role of sympathy and compassion in stimulating moral reflection and behaviour (Collier, 2010). He argued that the motivation to act ethically is provoked by emotion not just rational judgement. This position found footing in later critiques of rule-based normative ethical theory presented by MacIntyre (1984) and Nussbaum (1986), through a renewed interest in the Aristotelian concepts of moral character and virtue. Proponents argue that the role of moral philosophy is not to judge the value of specific acts *per se*, but rather to explore the fundamental disposition of the individual—their social qualities and mental faculties. As Nussbaum stressed, this involves an artistic component of ethical decision-making. Feeling, affect and emotion enhance our ability to understand moral situations and these elements shape our moral character. Vallor (2016) similarly asserts that creative thinking is essential to navigate the emerging worlds that technological innovation creates.

Johnson (1993) argues that to follow a model of ethics that just applies rules would ignore a key psychological component of morality—that of imagination. Across all forms of social interaction, abstract conceptualisation and reasoning, human beings apply imaginative thinking. How individuals interpret the world, and the right course of action, depends upon linguistic, cognitive and imaginative structures. These might include images, words (e.g. metaphors, discourses and narratives) and other aesthetic and sensory schema. As Johnson (1993) continues, our cognitive processes of moral reasoning are imaginatively structured. We require imagination in order to discern what is morally relevant in any given social situation; we must understand empathically how others experience things and must envision the full range of possibilities for action. Such thinking is important, not least because there is evidence of declining empathy within the public overall (Konrath et al., 2011; Terry & Cain, 2016). It is the combination of imagination and empathy that is important; empathy allows us to position ourselves as 'others', to connect with their emotions and physical circumstances, but moral imagination allows us to think beyond immediate interpersonal relationships towards a broader, abstract and hypothetical 'other'. Moral imagination allows us to consider our actions to broader society in a more general sense.

Moral imagination is the capacity to discover and evaluate possibilities beyond those dictated by immediate circumstance, going beyond the limitations of operative mental models, or sets of rules or rule-governed concerns (Werhane, 2015) such that it can stimulate and moderate empathic response to the plight of others, unconstrained by habitual thinking of what is right and wrong. It involves perceiving the interrelated norms, social roles and relationships of a situation, and the ability to envision and evaluate new mental models, new possibilities, to reframe ethical dilemmas and create new solutions in ways that are novel and justifiable. It also, therefore, becomes a means to correct *moral failures* that occur in decision-environments that commonly lead to defective choices. Notably these moral failures can occur when we rely upon empathy to make judgements, as discussed previously, proximity and similarity will often lead us to prioritise empathic judgements towards those that we see as closest and most similar to ourselves. It is the combination of empathy and moral imagination that encourages individuals to undergo a thorough examination of the ethical elements of a decision (Moberg & Seabright, 2000), by projecting one's empathic responses into situations (and indeed to people) that are unfamiliar. What is important here, is that VR has the unique capacity to stimulate and facilitate this process of moral imagination through immersive sensory stimulation, visualisation and interaction in virtual spaces. VR design is necessarily a function of imaginative, aesthetic and sensory components. Virtual worlds are limited only by the imagination of the designer and (to a varying extent) the user, and so such imaginative spaces created can unleash the moral imagination in unforeseen ways and reduce the burden upon human cognitive processes of imagination.

Pragmatism and Moral Imagination

Philosophical *pragmatism* has given impetus to re-examining the role of imagination in ethical reflection and practice (Radder, 2004; Schmidt-Felzmann, 2003; James, 1907). Pragmatism is less a systemic theory than a particular series of theses; "...which can be and were argued very differently by different philosophers with different concerns" (Grey, 1998, p. 255); see also (Putnam, 1994). Though diverse in approach, James neatly characterised pragmatism as:

> The attitude of looking away from first things, principles, categories, supposed necessities; and of looking towards last things, fruits, consequences, facts. (James, 1976, p. 55)

Pragmatism has seen a resurgence within philosophical thought in recent decades. Classical pragmatism is commonly understood as a uniquely North American tradition in philosophy and is notable for its relative absence from mainstream thought in the mid-twentieth century. The original proponents, the 'godfathers' of pragmatism: Charles Sanders Pierce, William James, George Herbert Mead and John Dewey (Wiener, 1974), were a group of intellectuals with diverse perspectives and research foci, though each had extensive influence on American and later international philosophy (Dalcourt, 1983). Bernstein (1992) argues, that despite their differences, the family resemblance between them is a persistent questioning of the idea that philosophy (or indeed any formal enquiry) rests upon fixed ontological foundations which can be known with certainty. Pragmatism presupposes that whilst philosophy should eschew commitments to foundational claims, it must also avoid descent into scepticism or pure relativism. Foundationalism and ethical monism are rejected on the grounds that preoccupation with general and abstract truths is counterproductive, in the sense that it distracts attention from concrete problems and conflicts tied to particular times, places and actors. A pragmatic ethics is premised therefore on the argument that ethical decision-making requires flexibility and context-sensitivity in order to be successful (Cherryholmes, 1999; Hickman, 2001; Michael, 2003; Keulartz et al., 2002). Pragmatic ethics can therefore be summarised by a commitment to the fallibility of enquiry, such that knowledge claims are open to potential criticism, and that the usefulness, workability and practicality of ideas are core criteria of their value. Pierce's maxim is simply:

> Consider the practical effects of the objects of your conception. Then, your conception of those effects is the whole of your conception of the object. (Peirce & Dewey, 2017)

Pragmatism sits within a consequentialist philosophical tradition, whereby the applicability of theory to *action* is the primary criterion of ethical validity (Festenstein, 1997), and where reflection upon such action opens the way to new insight (Parker, 1996). We can understand pragmatism therefore as a means of clarifying one's position through focus upon the end point of moral reasoning and thus it is a *method* rather than a specific doctrine (Dalcourt, 1983). Pragmatism-as-method rejects the view that human concepts and intellect can (solely and accurately) represent reality, and therefore stands in opposition to positivism and rationalism; asserting that only through the struggle of intelligent organisms with the surrounding environment can theories acquire significance, and it is only through a theory's success in this struggle that it *becomes* true (Peirce, 1982; Dewey, 1982; James, 1976, 1978). In short, as Rorty states: "pragmatists think that if something makes no difference to practice, it should make no difference to philosophy" (Rorty, 1995, p. 281).

John Dewey and Moral Imagination

Ethical tools are closely aligned to the pragmatist tradition in philosophy (Cotton, 2013); and they are often evaluated for their efficacy in assisting non-specialist actors in making good quality ethical decisions without recourse to substantive theory. This book builds upon the insights of John Dewey, using his philosophical works to develop a VR-based tool that draws together moral imagination and empathy within a technologically mediated system of ethical decision-making. Specifically John Dewey's work in "constructing the good" in ethical decision-making (McVea, 2007) through a focus upon empathy, creativity and the process described by Dewey as *dramatic rehearsal* (Collier, 2006) is used here to form the conceptual basis for a VR ethical tool discussed in the remainder of the book.

Dewey wrote extensively on ethics. Commentators note that his moral philosophy is fundamentally aesthetic and holistic in scope (Hamington, 2010). Dewey (1922, p. 164) states:

> For what is moral theory but a more conscious and systematic raising of the question which occupies the mind of anyone who in the face of moral conflict and doubt seeks a way out through reflection?

Dewey lived from 1859 to 1952, experiencing the Civil War in the United States, both World Wars, and the Cold War. The dramatic social, technological and cultural change from urbanisation, industrialisation and global conflict was reflected in his philosophy. Dewey believed that traditional moral norms, principles and philosophical traditions were insufficient for coping with the problems raised by dramatic transformations within society. He argued that ethical enquiry must involve reflective intelligence—bringing together thought and action. One must revise one's judgements in light of the consequences of acting upon them, such that reflection and action work in harmony with one another (Anderson, 2018). Dewey's ethics was fundamentally methodological—value judgements were tools to direct conduct, tested through practice. Moral progress is made by reflectively revising our judgements through deliberation and then aligning our habits through these reflective judgements. Dewey's ethics is therefore naturalistic and grounded in social psychology rather than in external or metaphysical justification.

As a method of ethics (or ethical tool, in this context) Dewey posits *dramatic rehearsal* to describe a type of deliberation for when people find themselves in indeterminate situations—when it is not clear how to act, what to value, or which ends to pursue. Caspary (2006) explains that moral deliberation for Dewey is dramatic in four ways: with concern for *character*, its concern for *plot*, its *difference from utilitarian* approaches and its *openness to unexpected circumstances*. Dewey recognised that people often feel blocked in their actions because existing routines, habits, norms, values, roles and responsibilities are destabilised. For example, novelties introduced by new developments in science and technology often emerge before new ethical norms about their governance and use become established (Krabbenborg, 2013). VR is a critical example of this phenomenon. When individuals encounter new forms of social interaction within virtual environments where common rules and norms of behaviour are less easily enforced, this can lead to the challenges outlined in Chaps. 1 and 2—individuals may over-share their personal experiences, online bullying and anonymous abuse may occur, and users may be less (or more) empathic than they would otherwise be in face-to-face offline interactions. The habitual social rules and practices are destabilised by the technology, leaving individuals in an uncertain position. Under these circumstances they are required to 'search' for new moral habits. In virtual environments, as in many new and unfamiliar forms of social conduct, the outcomes and moral values at stake are unclear. Dewey argues therefore for dramatic rehearsal as the 'work of discovery' (Dewey, 1932) that takes place—an attempt to find out, by enquiry, imagination and experimentation, what is at stake, which ends to pursue and what to value. By doing so, new ethical norms are explored and established. Rather than committing to and applying a specific normative ethical theory and then acting in accordance with it, Dewey argues that individuals must actively use their imagination to rehearse and evaluate a variety of responses and potential outcomes (Fesmire, 2003).

Though Dewey was concerned with the methodology of ethics, Caspary (2006) notes a sense of frustration that he did not provide a step-by-step process outlining the stages of how to turn dramatic rehearsal into a decision-procedure, nor its application to the working through of specific moral dilemmas. Fesmire (2003) argues that across Dewey's writings there are four modes of dramatic rehearsal, namely: *dialogue*, the *visualisation of results*, the *visualisation of performance*, and *imagination of possible criticism*. Krabbenborg (2013) outlines dramatic rehearsal as an interactive process. She states that the first phase of dramatic rehearsal is to *transform an indeterminate situation into a problematic situation*—this is a process of enquiry and the articulation of practical challenges presented by an ethical dilemma. Dewey emphasises that problems do not exist prior to an enquiry—by judging that something is a problem one then judges how it is to be defined and resolved. The second phase involves *developing hypotheses about potential solutions*, thinking through the possible consequences of executing particular lines of action. The third phase is *rehearsal*, an interactive and imaginative process through which the potential consequences of each line of action are reflected upon from the perspective of those affected by the action. In the final phase is the *experimental testing of the hypothesis* that emerged as the best solution (and potentially a new dramatic rehearsal). The best solution should incorporate as many issues as possible that were uncovered during the preceding steps. Dramatic rehearsal involves deductive reasoning about the future—testing hypotheses about how lines of action lead to consequence, but this is also an imaginative process—the futures are uncertain and require interpretation through dramatisation to make them visible in the mind's eye.

Though grounding for both the deductive and imaginative elements of dramatic rehearsal is moral empathy. Dewey argued that caring for others is based upon an empathic response. Dramatic rehearsal is the method or thought procedure through which empathic ethical decision-making is grounded and contextualised—one first encounters a moral problem, imagines the courses of action and actors involved, empathises with them in the context of different courses of moral action, and then 'rehearses' the outcomes of the different lines of action available, until an action is decided upon on the basis of a process of deliberation about the moral acceptability of the outcome. Fesmire (2003) argues that dramatic rehearsal is not simply a rule-based procedure for choosing the right course of action. It is better understood as having an organic unity with Dewey's pragmatic theories of habit, social self-belief and enquiry, with the thread of imagination running throughout. To Dewey, the process of moral deliberation involves an individual or collective *hunt* for the different ways to clarify objectives, scope alternatives and imagine how we can take part in actions that will create future outcomes. "Hunting for futures" continues as a process of rehearsal until one is ready to act—a point where the interests, needs and other factors of the situation harmonise such that the moral deliberation is felt to be complete.

Fundamentally, dramatic rehearsal relies upon *deliberation*. When presented with a problem one must first imagine what that potential outcomes will be, and then 'try on' one or other of the ends (i.e. to imagine oneself actually doing or being affected by them). This is necessary as to provide context and structure to one's understanding of the consequences of moral action; as Misak (2000, p. 52) argues:

> The practice of moral deliberation is responsive to experience, reason, argument, and thought experiments… [this notion of] responsiveness is part of what it is to make a moral decision and part of what it is to try to live a moral life.

Dewey argues that deliberation on the consequences of moral action is, in essence, a type of make-believe—it is this sense that ethics is a 'dramatic' action. However, the dramatic mechanism by which to achieve the deliberation of future outcomes might differ from case to case, or from person to person. For example, deliberation may occur through dialogue between concerned parties, through an individual visualising certain results or practising doing different things. There is flexibility in how the rehearsal part takes place but there are commonalities of approach. First, one must have empathy, such that they can put themselves in the shoes of another person (or people) that might be affected by one's actions. Second, one must imagine what the consequences of one's actions would be to the individual. Third, one engages in exploring a range of different scenarios under conditions of deliberation, in the process of dramatic rehearsal. This process extends the temporal horizons of moral deliberation, and it takes time to play out imaginative possibilities, moral rules and the consequential calculations involved. For Dewey, ethical decision-making includes concern not only for the immediate resolution of a critical dilemma, but also for long-term relationship-building, the alignment social practices and habits with philosophical reflection, and the incremental trial and error of ethics as an experiment in imaginative practice.

Conclusion: Dramatic Rehearsal as Ethical Tool

The development of ethical tools necessarily involves an imaginative process of engagement between thought and action. The stimulation of empathic engagement is a necessary precondition, though is insufficient to ensure good moral conduct. Empathy is a human trait that allows us to connect with the thoughts and feelings of others, but without guidance our empathy can lead to moral mistakes—such as biases of similarity through proximity and identity. The empathic bias towards the in-group needs to be corrected if an ethical decision is to be made. It is in this way that Dewey's concept of dramatic rehearsal has considerable value. Dramatic rehearsal guides one's empathic response through a 'hunting phase' of problem identification, followed by imaginative engagement with potential lines of action and their consequences, and the consideration of outcomes for a range of different people, not just the individual which one is empathising with. It is in this way that dramatic rehearsal structures our empathy and provides what Bloom (2017a) calls *rational compassion*. Rather than being caged by our biases and structured moral habits, dramatic rehearsal as ethical decision-making method encourages greater reflection, deliberation and connection to personal value change in practice as well as in thought.

Dramatic rehearsal provides a conceptual structure that links moral imagination with empathy. However, as an ethical tool it is important to note that the practical steps are open to methodological interpretation and insight from contemporary collaborative and deliberative methods in applied philosophy and the social sciences. For some, such as Fesmire (1994), artistic methods are highly valued; he argues that one must adopt the attitude of the moral artist, practising dramatic rehearsal to the creation of works of art that engage our imagination (see also Spencer, 2013). Others, such as Hamington, recommend that character acting or method acting can exercise one's imaginative capacities through dramatic embodiment as a means to develop moral empathy. Cotton (2013) meanwhile brings in the methodology of future studies, namely the backcasting approach, in which one uses brainstorming, concept mapping and group deliberation to imagine idealised moral futures, and then work backwards from the future to the present identifying practical steps to encourage that future to emerge. There is great diversity in the approaches used to achieve dramatic rehearsal. In keeping with the pragmatist commitment to understanding philosophical concepts through their application to practice, in the subsequent chapters I discuss how the steps of dramatic rehearsal can be integrated into a VR-based ethical tool. In Chap. 5, I present a series of vignettes illustrating how VR is applied in different contexts to stimulate imaginative and empathic ethical engagement with a range of social problems. In the concluding Chap. 6, I bring together a discussion of dramatic rehearsal and the analysis of practical cases to distil specific features for the design of a VR programme.

References

Almond, B. (1988). Women's right: Reflections on ethics and gender. In M. Griffiths & M. Whitford (Eds.), *Feminist perspectives in philosophy*. Indiana University Press.

Anderson, E. (2018). Dewey's moral philosophy. In *Stanford encyclopedia of philosophy* [Online]. Version. https://plato.stanford.edu/entries/dewey-moral/

Baggini, J., & Fosl, P. S. (2007). *The ethics toolkit: A compendium of ethical concepts and methods*. Wiley-Blackwell.
Baron-Cohen, S. (2012). *The science of evil: On empathy and the origins of cruelty*. Basic Books.
Battaly, H. D. (2011). Is empathy a virtue? In A. Coplan & P. Goldie (Eds.), *Empathy: Philosophical and psychological perspectives* (pp. 277–301). Oxford University Press.
Beauchamp, T. L., & Childress, J. F. (2001). *Principles of biomedical ethics* (5th ed.). Oxford University Press.
Bernstein, R. J. (1992). The resurgence of pragmatism. *Social Research, 59*, 813–840.
Bloom, P. (2017a). *Against empathy: The case for rational compassion*. Random House.
Bloom, P. (2017b). Empathy and its discontents. *Trends in Cognitive Sciences, 21*(1), 24–31.
Brandt, R. B. (1959). *Ethical theory: The problems of normative and critical ethics*. Prentice-Hall.
Brennan, S. (1999). A survey of recent work in feminist ethics. *Ethics, 109*(4), 858–893.
Brennan, S. (Ed.). (2002). *Feminist moral philosophy*. University of Calgary Press.
Caspary, W. R. (2006). Dewey and Sartre on ethical decisions: Dramatic rehearsal versus radical choice. *Transactions of the Charles S. Peirce Society, 42*(3), 367–393.
Cherryholmes, C. (1999). *Reading pragmatism*. Teachers College Press.
Chismar, D. (1988). Empathy and sympathy: The important difference. *The Journal of Value Inquiry, 22*(4), 257–266.
Cialdini, R. B., Brown, S. L., Lewis, B. P., Luce, C., & Neuberg, S. L. (1997). Reinterpreting the empathy-altruism relationship: When one into one equals oneness. *Journal of Personality and Social Psychology, 73*(3), 481.
Clohesy, A. M. (2013). *Politics of empathy: Ethics, solidarity, recognition*. Routledge.
Collier, J. (2006). The art of moral imagination: Ethics in the practice of architecture. *Journal of Business Ethics, 66*, 307–317.
Collier, M. (2010). Hume's theory of moral imagination. *History of Philosophy Quarterly, 27*(3), 255–273.
Cotton, M. (2013). Deliberating intergenerational environmental equity: A pragmatic, future studies approach. *Environmental Values, 22*(3), 317–337.
Dalcourt, G. J. (1983). *The methods of ethics*. University Press of America.
De Vignemont, F., & Singer, T. (2006). The empathic brain: How, when and why? *Trends in Cognitive Sciences, 10*(10), 435–441.
Decety, J., & Jackson, P. L. (2004). The functional architecture of human empathy. *Behavioral and Cognitive Neuroscience Reviews, 3*(2), 71–100.
Dewey, J. (1922). *Human Nature and Conduct*. Cambridge: Cambridge University Press.

Dewey, J. (1932). Ethics. IN J. Boydston (Ed.), *John Dewey: The Later Works, 1925–1953, 17.* Carbondale IL: Southern Illinois University Press.

Dewey, J. (1982). The pattern of inquiry. In H. S. Thayer (Ed.), *Pragmatism: The classic writings.* Hackett.

Doppelt, G. (2002). Can traditional ethical theory meet the challenges of feminism, multicuturalism, and environmentalism? *Journal of Ethics, 6*, 383–405.

Dunn, R. (2004). Moral psychology and expressivism. *European Journal of Philosophy, 12*(2), 178–198.

Ferreira, M. J. (1994). Hume and imagination: Sympathy and 'the other'. *International Philosophical Quarterly, 34*(1), 39–57.

Fesmire, S. (1994). Educating the moral artist: Dramatic rehearsal in moral education. *Studies in Philosophy and Education, 13*(3–4), 213–227.

Fesmire, S. (2003). *John Dewey and moral imagination: Pragmatism in ethics.* Indiana University Press.

Festenstein, M. (1997). *Pragmatism and political theory.* Polity.

Garner, R. T., & Rosen, B. (1967). *Moral philosophy: A systematic introduction to normative ethics and meta-ethics.* Macmillan.

Gilligan, C. (1982). *In a different voice: Psychological theory and women's development.* Harvard University Press.

Grey, T. (1998). Freestanding legal pragmatism. In M. Dickstein (Ed.), *The Revival of pragmatism: New essays on social thought, law, and culture* (pp. 254–274). Duke University Press.

Haidt, J. (2001). The emotional dog and its rational tail: A social intuitionist approach to moral judgment. *Psychological Review, 108*, 814–834.

Haidt, J. (2003). The moral emotions. In R. J. Davidson, K. R. Scherer, & H. H. Goldsmith (Eds.), *Handbook of affective sciences* (pp. 852–870). Oxford University Press.

Hamington, M. (2010). Care ethics, John Dewey's 'dramatic rehearsal,' and moral education. *Philosophy of Education Archive*, pp. 121–128.

Haney, K. (1994). Empathy and ethics. *Southwest Philosophy Review, 10*(1), 57–65.

Heath, E. (1995). The commerce of sympathy: Adam Smith on the emergence of morals. *Journal of the History of Philosophy, 33*(3), 447–466.

Hickman, L. (2001). *Philosophical tools for technological culture: Putting pragmatism to work.* Indiana University Press.

Hoffman, H. G., Patterson, D. R., & Carrougher, G. J. (2000). Use of virtual reality for adjunctive treatment of adult burn pain during physical therapy: A controlled study. *The Clinical Journal of Pain, 16*(3), 244–250.

Hume, D. (1739). *Treatise on human nature: Of virtue and vice in general.* Oxford University Press.

Husserl, E. (1931). *Cartesian meditations.* Martinus Nijhoff.

Hutcheson, F. (1725). *An essay on the nature and conduct of the passions and affections. With illustrations on the moral sense. By the author of the inquiry into the*

original of our ideas of beauty and virtue. London: J. and J. Knapton, John Darby, Thomas Osborne, Jauton Gilliver, John Crownfield.
James, W. (1907). *Pragmatism: A new name for some old ways of thinking*. Longmans, Green and Co.
James, W. (1976). *Pragmatism*. Harvard University Press.
James, W. (1978). The meaning of truth. In F. H. Buckhardt (Ed.), *Essays in philosophy: The works of William James*. Harvard University Press.
Johnson, M. (1993). *Moral Imagination: Implications of cognitive science for ethics*. University of Chicago Press.
Keulartz, J., Korthals, M., Schermer, M., & Swierstra, T. E. (Eds.). (2002). *Pragmatist ethics for a technological culture*. Kluwer.
Kohlberg, L. (1984). *The psychology of moral development : The nature and validity of moral stages*. Harper & Row.
Konrath, S. H., O'Brien, E. H., & Hsing, C. (2011). Changes in dispositional empathy in American college students over time: A meta-analysis. *Personality and Social Psychology Review, 15*(2), 180–198.
Krabbenborg, L. (2013). Dramatic rehearsal on the societal embedding of the lithium chip. In S. van der Burg & T. Swierstra (Eds.), *Ethics on the laboratory floor* (pp. 168–187). Springer.
Lekan, T. (2006). Pragmatist metaethics: Moral theory as deliberative practice. *Southern Journal of Philosophy, 44*(2), 253–272.
Lipps, T. (1903). *Aesthetik*. Voss Verlag.
MacIntyre, A. (1984). *After virtue: A study in moral theory*. University of Notre Dame Press.
McCracken, J., & Shaw, B. (1995). Virtue ethics and contractarianism: Towards a reconciliation. *Business Ethics Quarterly*, pp. 297–312.
McVea, J. F. (2007). Constructing good decisions in ethically charged situations: The role of dramatic rehearsal. *Journal of Business Ethics, 70*, 375–390.
Michael, M. A. (2003). What's in a name? Pragmatism, essentialism, and environmental ethics. *Environmental Values, 12*, 361–379.
Misak, C. (2000). *Truth, politics, morality*. Routledge.
Moberg, D., & Seabright, M. (2000). The development of moral imagination. *Business Ethics Quarterly, 10*, 845–884.
Nussbaum, M. (1986). *The fragility of goodness: Luck and ethics in Greek tragedy and philosophy*. Cambridge University Press.
Nussbaum, M. (2001). *Upheavals of thought: The intelligence of the emotions*. Cambridge University Press.
Parker, K. A. (1996). Pragmatism and environmental thought. In A. Light & E. Katz (Eds.), *Environmental pragmatism*. Routledge.
Peirce, C. S. (1982). Definition and description of pragmatism. In H. S. Thayer (Ed.), *Pragmatism: The classic writings*. Hackett.
Peirce, C. S., & Dewey, J. (2017). How to make our ideas clear. In C. Pierce (Ed.), *Chance, love, and logic* (pp. 32–60). Routledge.

Pellizzoni, L. (2012). Strong will in a messy world. Ethics and the government of technoscience. *NanoEthics, 6*(3), 257–272.
Porter, E. (1999). *Feminist perspectives on ethics.* Longman.
Putnam, H. (1994). *Words and life.* Harvard University Press.
Radder, H. (2004). Pragmatism, ethics, and technology'. *Techné: Science, Technology & Human Values, 7*(3), 10–18.
Rorty, R. (1995). Is truth a goal of inquiry? Davidson vs. Wright. *Philosophical Quarterly, 45*(189), 281–300.
Schmidt-Felzmann, H. (2003). Pragmatic principles—Methodological pragmatism in the principle-based approach to bioethics. *Journal of Medicine and Philosophy, 28*(5–6), 581–596.
Schumann, K., Zaki, J., & Dweck, C. S. (2014). Addressing the empathy deficit: Beliefs about the malleability of empathy predict effortful responses when empathy is challenging. *Journal of Personality and Social Psychology, 107*(3), 475–493.
Shrage, L. (1994). Interpretative ethics, cultural relativism and feminist theory. In L. Shrage (Ed.), *Moral dilemmas of feminism* (pp. 162–184). Routledge.
Sinnott-Armstrong, W. (1987). Moral realisms and moral dilemmas. *The Journal of Philosophy, 84*(263–276).
Slote, M. (2007). *The ethics of care and empathy.* Routledge.
Slote, M. (2017). The many faces of empathy. *Philosophia, 45*(3), 843–855.
Songhorian, S. (2019). The contribution of empathy to ethics. *International Journal of Philosophical Studies, 27*(2), 244–264.
Spencer, A. R. (2013). The dialogues as dramatic rehearsal: Plato's republic and the moral accounting metaphor. *The Pluralist, 8*(2), 26–35.
Stein, E. (1917). *On the problem of empathy.* ICS Publishers.
Sumner, L. W. (1967). Normative ethics and metaethics. *Ethics, 77*(2), 95–106.
Terry, C., & Cain, J. (2016). The emerging issue of digital empathy. *American Journal of Pharmaceutical Education, 80*(4), 58.
Vallor, S. (2016). *Technology and the virtues: A philosophical guide to a future worth wanting.* Oxford University Press.
Walker, M. U. (1989). Moral understandings: A feminist study in ethics. *Hypatia, 4*(2), 15–28.
Werhane, P. H. (2015). Moral imagination. *Wiley encyclopedia of management,* 2(1–2), John Wiley & Sons, Ltd. https://doi.org/10.1002/9781118785317.weom020036
Wiener, P. P. (1974). Pragmatism. In P. P. Wiener (Ed.), *The dictionary of the history of ideas: Studies of selected pivotal ideas* (pp. 551–570). Charles Scribner's Sons.
Zahn-Waxler, C., Hollenbeck, B., & Radke-Yarrow, M. (1985). The origins of empathy and altruism. In M. W. Fox & L. Mickley (Eds.), *Advances in animal welfare science 1984* (pp. 21–41). Springer.

CHAPTER 5

Virtual Reality as Ethical Tool

Abstract This chapter examines the application of virtual reality (VR) to prosocial behaviour change, news reporting, art and social justice campaigning. A range of cases are considered, including the immersive journalism of Nonny De La Peña on abortion rights, poverty and suspect interrogation; the awareness raising/fundraising efforts of The United Nations Virtual Reality (UNVR) Series, including the acclaimed "Clouds Over Sidra" VR film, and applications to art and gaming through platforms including: "The Machine to Be Another (TMTBA)", and "A Breath-taking Journey" (ABTJ). The capacity of these works to stimulate empathy is critically assessed and lessons drawn from their design for the construction of a VR ethical tool.

Keywords Prosocial behaviour • United Nations Virtual Reality Series • Embodiment • Empathy-arousal

Introduction

Previous chapters discussed the role that virtual reality (VR) plays in creating immersive environments through visual and haptic feedback, motion-sensitive visual display, motion-tracking and user interaction. It is for this reason that the technology has become so closely associated with gaming, fantasy and escapism. It is portrayed in media and advertising as a way for

M. Cotton, *Virtual Reality, Empathy and Ethics*,
https://doi.org/10.1007/978-3-030-72907-3_5

individuals to experience more-than-real encounters through imaginative design. However, the association with gaming has conversely led to the practical applications of VR for other beneficial purposes being overlooked by the mainstream media, educational and training institutions, and professional organisations. In short, VR is categorised as a recreational activity in the popular imagination, and so is often dismissed as a serious tool for prosocial change.

As mentioned in Chap. 2, certain sectors are beginning to buck this trend, harnessing VR primarily in training, education and therapeutic environments. As discussed in chapter the advent of cheap, high-quality head-mounted displays (at the time of writing Facebook's Oculus™ or HTC Valve™ headsets are sold within consumer price brackets) or even lower-cost adaptations of mobile phone devices (such as Google Cardboard VR™) are moving VR from niche technology to broader mainstream appeal. As a result, there is growing research and development interest in the application of VR not only to training but to more overtly prosocial ends. VR is being used in journalism, fundraising, art and social justice movements to foster public empathy for the lived experience of marginalised, traumatised or oppressed communities—including disaster victims, prisoners, refugees and children suffering extreme poverty and violence. The hope is that by exposing socially privileged people to the experience of marginalised groups through the immersive sensory experience of VR, this will, in turn, foster greater empathy for the plight of these peoples and longer-term attitudinal and behaviour change. The following section outlines a range of VR applications to prosocial justice and behaviour change causes, followed by a critical analysis of their effectiveness in stimulating empathy, good lessons drawn to the development of an ethical tool discussed in Chap. 6.

Virtual Reality and Prosocial Engagement

A number of important prosocial justice and public interest VR programmes have emerged in recent years. The following section discusses a range of different causes, VR technology platforms, types and applications that have emerged in recent years. There are four platforms that are of critical interest:

1. The work of Nonny De La Peña
2. The United Nations Virtual Reality Series (UNVRS)

3. The Machine to Be Another (TMTBA)
4. A Breath-taking Journey (ABTJ)

These artistic, journalistic and gaming technology platforms are used in different ways to meet at least one of three primary objectives. The first is to *stimulate a sense of immersion*, the second is to *stimulate a sense of embodiment*, and the third is to be *persuasive*. It is the combination of these three elements when used together that stimulates an empathic response, and in each case the promoters of the technology aim to compel the user to undergo long-term value and behaviour change. The immersive and embodiment capabilities of VR to stimulate empathy work by drawing a connection between the individual VR user and a broader ethical dilemma. As VR film director Chris Milk states, VR technology creates the "ultimate empathy machine": a system that can facilitate an empathic connection in a unique way. In this chapter, I critically evaluate this claim with reference to the practical applications of VR promoted by artists, documentary film-makers and non-governmental institutions. From this evaluation I draw a series of general principles derived for the development of an empathic and imaginative ethical tool founded in dramatic rehearsal, and then in the final concluding chapter I present a programmatic proposal for the development of new VR programmes for ethical decision-making that brings these different elements together.

Immersion, Embodiment and Persuasion

VR as a communicative medium provides new opportunities for artists, film-makers and journalists to reach new audiences by stimulating users' sensory engagement with narrative. In journalism, VR has been described as a radically different form of presentation, one that providers a sensory experience of reality unmatched by media consumption through window-on-a-world 2D film, print or televised address. As Sirkkunen, E. and Uskali, T. (2019) argue, VR provides a sense of immediate presence within a story—unmatched by print or audio-visual media. The sense of immersion produced by VR is unique in simulating "temporally or spatially distant real environments" (Steuer, 1994), which in turn, makes the experience of narrative more concrete to user.

Immersion and presence are relevant to the social psychological phenomenon of *construal levels*—that the further away something is perceived by the individual, it is in turn, construed in more abstract terms (Förster et al., 2004). Conversely, the closer something is perceived, the more concrete it appears. Research on construal levels has been applied to range of different phenomena, including consumer choice, estimations of probability, financial decision-making, perceptions of climate change, or social group differences (Trope & Liberman, 2010). The relative abstraction of the problem based upon its perceived proximity is the key barrier for empathic engagement with the ethics of the situation. For many important social phenomena that involve ethical choices and values seem distant from the everyday lives of those in a position to ameliorate suffering and bring about positive social change. Providing financial aid to disaster victims, food for the poor, better treatment of prisoners, safe harbour for refugees, or fair treatment of racially oppressed or economically marginalised communities is abstract and distant for those in a position of wealth, safety, privilege and power. The insulating effect of wealth and privilege means that the plight of marginalised groups is construed as abstract due to a lack of personal experience, shared understanding and geographic separation. Encounters between different social groups on opposite ends of socio-economic spectrum either do not occur, or when they do, conflict emerges due to abstract perceptions of difference that heighten hostilities. From a VR-journalism perspective, reducing the temporal, spatial and social distance mediated through technology allows us to create stronger empathic bonds by making the experiences of others more concrete, thus changing the construal level of the problem for the party in a position of power to enact change.

There are a number of emergent examples of VR journalism attempts to improve the immersion experienced by the audience within a virtual environment. Immersive journalism employs 360° video within the journalistic production method using a vision system based upon stereoscopy within a VR headset. The method of 360° video had been used by real estate developers, architects, and in education and advertising. However, when transferred to journalism, the argument is that one can experience what it is like to be in the scenario in which the news takes place. VR journalism provides access to new spatial points of view that facilitate comprehensive visual narration. The aim is to increase user understanding and sensitivity towards the reality depicted within the film, based upon the principle that human knowledge increases when first-person experiences are available. The participant becomes part of the story within immersive journalism, either by viewing the story as oneself, or else through the perspective of a character within the journalistic narrative (Benítez de Gracia et al., 2019). This character can either be the subject of the narrative or a bystander to the central story. The underlying ethos of the immersive journalism style is that by visiting the virtual space as oneself or as a subject in the narrative, the participant connects to the real event through unprecedented access to the sights and sounds that occur within the story, and thus captures a deeper sense of the affective and emotional states that the news story elicits.

The Work of Nonny De la Peña

Immersive journalism using VR is pioneered by journalism scholar and documentary film-maker Nonny De la Peña. De la Peña became known for her collaboration with Peggy Weil on the VR experience 'Gone Gitmo' that digitally reconstructs the interrogation of detainee 063, Muhammed al-Qahtani in Guantánamo Bay prison in 2002 and 2003. Al-Qahtani was imprisoned, hooded and constrained in stress positions through interrogation techniques used at that time on suspected terrorism suspects. The film came out in 2009 at the beginning of the Obama Administration, during a period in which political debate over the closing of Guantánamo Bay during the War on Terror had become heated. The virtual experience through a head-mounted display brings together a combination of news reports on detainees being held for extended periods in stress positions, and user experience of the prison environment through a computer-generated reconstruction in Second Life™. The combination of news journalism and the illusory transformation created by immersive virtual reality was constructed in such a way that the user can experience, by proxy, the body of another under stressful conditions. The topic is a powerful one for this type of immersive journalism, simply because access to Guantánamo Bay prison is so severely limited that the broader civil society political debate on the ethics of justice in the context of the War on Terror, extraordinary rendition, imprisonment without trial, and interrogation bordering on torture, took place among civil society actors that have no direct information or experience of the situation at hand. VR is a powerful tool for placing people into restricted physical spaces, as well as into the perspective of people from very different backgrounds and experiences. De la Peña, in collaboration with scientists Mel Slater and Maria Sanchez-Vives, argued that immersive experience provides an illusory transformation of the physical body such that one perceptually enters the body of another (De la Peña et al., 2010).

De la Peña continued development of immersive journalistic techniques through VR film that combined documentary evidence and 3-D modelling to simulate the experience of people at a food bank in Los Angeles witnessing a man going into diabetic shock. The *Hunger in Los Angeles* project centres upon the intersection of poverty and medical crisis. The 3D environment recreates the experience of those waiting in the queue, bystanders and those that intervene. The user of the VR headset becomes part of that crowd. Perception within the re-created 3D environment was immersive to the point that users reportedly stepped around virtual characters so as to avoid collision. Similarly, in collaboration with creators Brad Lichtenstein, Jeff Fitzsimmons and the organisation Planned Parenthood, De la Peña created the VR experience *Across the Line*, detailing the experiences of women seeking planned pregnancy termination, and the health centre staff at Planned Parenthood facilities. The virtual experience moves between 360° video and computer-generated imaging, where users shift between real-life sounds and images of protests outside healthcare facilities, and a game-like environment where VR users must navigate a 3D computer-generated environment, weaving in and around protesters shouting shaming and stigmatising remarks. The aim of the film is to capture the experience of women trying to access these health services amidst the febrile political atmosphere surrounding reproductive health services in the United States.

The Hunger in Los Angeles (HILA) and Across the Line (ATL) experiences are based upon real-life events, but the technology blends real film, documentation and news reporting with computer-generated content (though the computer-generated content reconstructs real places, rather than invents them). The tacit connection between real places and people, the lived experience of real events, and the computer-generated simulation may help to explain why VR users behave as if they are within a real environment, avoiding collision with objects and people, and navigating the space in a realistic manner. Even though characters are digitally created using 3D modelling rather than through 360° film, the connection between the digital simulation and the reality is drawn within peoples' minds such that an empathic engagement with the situation can occur. It is here, therefore, that the three elements of VR come together, for example in Gone Gitmo the immersion of the user within the virtual environment of Guantánamo Bay, the textual news reporting of detainee abuse, and the illusory embodiment of the detainee are more provocative and affecting than any of the individual elements on their own. It is the combination of different forms of sensory stimulation, information and immersion that turns the technology into a form of *persuasive* media. The embodiment and immersion create what I would call a body-linked politics, sensitive to interpretive flexibility by the user. The experience acts as an antagonist, stimulating not only empathy with the detainee/patient/victim, but connecting bodily sensation to the broader civic dialogue around detention, torture, the War on Terror, poverty or abortion and so on. It is hoped by the creators that this will stimulate broader dialogue and social learning about the situation beyond just the user experience.

The United Nations VR Series

As part of the United Nations sustainable development goals action campaign, the UN launched the VR series. Their aim was to:

> ...bring the world's most pressing challenges home to decision makers and global citizens around the world, pushing the bounds of empathy The UN Virtual Reality Series shows the human story behind development challenges, allowing people with the power to make a difference have a deeper understanding of the everyday realities of those in danger of being left behind. (UNVR, 2020)

The prosocial motivation is clear in the UNVR Series. The programme is part of the Sustainable Development Goal Action Campaign (SDGAC) aimed at drawing attention to the voices of marginalised people across the world. The UNVR Series seeks to document everyday narratives from those experiencing traumas as a result of natural disasters, conflict, disease outbreaks, climate change or other human-driven crises. As in the immersive journalism model, the VR platforms use a combination of headsets and stereoscopic 360° film to promote the message. Unlike a journalistic endeavour that seeks to highlight these narratives for a general public audience, however, the UNVR Series is specifically focused upon political decision-makers, funders, humanitarian organisations, charitable foundations and intergovernmental agencies. The films serve to both raise awareness amongst networks of global development stakeholders, and to act as fundraising tools through empathy arousal.

At the time of writing there are 21 films on the UNVR roster. The majority were produced between 2015 and 2019. The films display a range of deeply affecting experiences of social and environmental crisis.

For example, the film *Waves Of Grace* by Vice Media explores the experiences of a woman called Decontee from Liberia during the largest Ebola outbreak in history. It charts her experiences within the community of West Point in Liberia, how she survived the deadly disease, the loss of her family, and consequently the stigmatisation she faced, and the challenges of her work in helping children orphaned during the crisis. Another film called *Ground Beneath Her* explores the experiences of 14-year-old Sabita, whose home was damaged in April 2015 by the 7.8 magnitude earthquake that killed 9000 people and destroyed more than half a million homes and buildings in Nepal. The film shows the daily struggles that she endures in trying to balance home and school life in the wake of the disaster recovery. Collectively, the film-making covers multiple countries and case study contexts within the UNVR series, though one of the most politically significant of these was the 2015 VR film: *Clouds Over Sidra* (hereafter COS).

COS is an eight-minute-long VR film developed by Chris Milk, and directed by Barry Pousman and Gabo Arora. The objective of the film was to show public audiences and key decision-makers the crisis developing in the Za'atari refugee camp in Jordan, home to (in 2015) 84,000 Syrian refugees fleeing conflict with the Assad regime. The film follows a 12-year-old girl, who lives in the camp along with her family. Locations in the film include her family's tent, a makeshift school, the gym, a bakery, computer lab and the football pitch with a rubble-strewn playing field. The immersive film documents life in the refugee camp against a background of growing cloud cover—which exerts an ominous presence. The influence and impact the film portrayed in the virtual medium raises important questions about the way that subjects portray, journalists capture, and consumers learn about news (Kool, 2016).

The political significance of COS was established after its showing at a conference organised by the World Economic Forum (WEF) at Davos in Switzerland. The WEF is a not-for-profit organisation with a self-stated commitment "to improving the state of the world by engaging business, political, academic, and other leaders of society to shape global, regional, and industry agendas" (The World Economic Forum, 2020). The aim of the screening at Davos was to raise funds for refugees and highlight the plight of those affected by the Syrian conflict. In this, it was highly successful. At the humanitarian pledging conference participants donated $3.8 billion ($2.5 billion) for Syrian refugees, surpassing the $2.3 billion total that was projected (Anderson, 2015). The power of VR in making the presence of the refugees' plight known to donors in particular is deeply significant. As Milk (2015) states:

> We have been referring to virtual reality as the *ultimate empathy machine* [emphasis added]... These films can help donors understand the everyday reality of ordinary people caught in the middle of conflict [and] can help form policies and raise funding.

As such, the UN Millennium Campaign (UNVR, 2020) described the project as helpful in raising funds and in bringing:

> ...the experience of vulnerable communities straight to decision-makers, thereby creating deeper empathy and understanding. This is in line with the UNMC's efforts to elevate the voices of those who often do not have a say, bringing people's voices directly into the decision-making process.

In COS the user of the VR technology can more easily relate to the experience and lived reality of the refugee plight through a combination of immersive journalism and a focus upon vulnerable people through narrative. The VR experience conveys the perspectives of the journalistic subjects through emotionally charged content (a combination of stark images and 'overheard' naturalistic conversations) which makes the stories more persuasive (De la Peña et al., 2010) and provokes a sense of moving beyond superficial image-based representations of Sidra's experience, towards an immersive sense of being alongside her and her classmates. The experience remains a visualised journalistic medium, and yet as Jones (2017) argues, this type of VR-led journalism nominally simulates the role of the reporter to the user of the technology through a type of proxy immersion in a news event.

Clouds Over Sidra is illustrative of the capacity of VR to assist in storytelling in four ways. The first is by the omission of the camera and hence the *erasure of the journalist* (reducing the sense of distance felt between the viewer and the subject of the media narrative; Kool, 2016). The second is the enhanced sense of engagement in the reality as sensory information supplements the intake of the narrative (Bailey et al., 2012), which according to *media richness theory* (Daft & Lengel, 1986) enhances the depth and quality of narrative information, such that a felt experience of *presence* is stimulated to create what could be termed an *immersive witness* (Nash, 2018). Third, by introducing interactive elements into virtually recreated scenarios this stimulates emotions that will influence action (Shin, 2017). The fourth is what I term ***deep visual representation***. The power of visual representation is important in a world in which we are "flooded with images and spectacles to the extent that it is superficially and depthlessly contrived" (Aitken & Craine, 2009). Deep visual representation is achieved in COS through the preceding three elements—erasure of the journalist, the enhancement of presence and the empathic driver for action to resolve the plight of Syrian refugees fleeing conflict in their country. The narrative is captivating in a way that 'clickbait' news headlines, social media feeds and rolling news are not.

What is also significant of deep visual representation is user autonomy. VR users can make decisions about what they see and hear about the story. The move from linear narrative (i.e. controlled communication from storyteller to receiver) to a branching narrative leads us to question what is gained by giving the receiver more freedom 'to explore' (Hodgkinson, 2016). The VR blurs the boundaries of production and viewing. Yet, unless there is significant interactivity within the narrative, this remains 'framed' for the most part by the reporter (and directors, editors and production companies). When we think about the application of COS-style VR to facilitating a process of ethical reflection, this issue of framing is significant. Without choice there is no internal deliberation that takes place about future worlds created by the actions in response to the witnessing of Sidra's experiences. If the VR user is passive, they are not required to make decisions about what to do next. They may seek to explore the narrative in order to learn more about the situation, but do not have direct involvement in the consequences of what happens next.

The makers of COS intended for decisions over what happens next to occur outside *of* the virtual realm. The VR was intended as an empathic stimulus which then carries over into the economic-political deliberation that occurs between aid organisations, non-governmental organisations (NGOs) and wealthy donors. There is also a broader framing effect in place. In the case of COS the VR film was funded by the UN, but also formed through collaboration with the South Korean technology giant Samsung. So, whilst the intention of the film-maker Chris Milk was clearly philanthropic, questions are raised as to whether much of the empathy it garners is truly a marketing ploy for the "distribution of the Samsung headset, the diplomacy of the UN, or the attention for Vrse?" (Kool, 2016). Each of the actors involved in the film-making, distribution and technology platform have their own intent and audience to influence. This in turn, influences the nature and quality of the ethical deliberation that takes place about the technology and the response to the refugee crisis itself.

The Machine to Be Another

Though immersive narrative has been utilised in documentary and journalistic reporting, the prosocial benefits of VR are also harnessed through artistic process. Of key interest is the Open-Source Art project called *The Machine to Be Another* (TMTBA). TMTBA is part art installation, part participatory research project and part psychological experiment. The stated aims of the project are to explore the relationships between human identity and empathy through digital experimentation. The project uses a bespoke VR technology to explore issues of embodiment and virtual body extension (BeAnotherLab, 2017). TMTBA aims to directly stimulate a sense of ***embodiment*** in the user. Unlike the COS which has a relatively passive (though immersive) sense of the user within the virtual space, TMTBA creates a type of performative VR, in which the user can see themselves in a different body and move around a virtual space with tactile feedback. Whereas COS encourages the user to think of themselves in the context of Sidra's home in the refugee camp, the aim of TMTBA is to trick the mind into imagining itself within another body, as a means allow one to 'step into the shoes of another'; as de Oliveira Musse et al. (2018) argue, the TMTBA produces an illusionary effect of a full body using another real body inside the VR environment (rather than an avatar or game character).

The system is built to stimulate a *body ownership illusion*. The sense of bodily ownership can be manipulated through changes to visual, tactile and proprioceptive environments—through which perceptions of bodily location, shape and size can be manipulated, thus tricking the brain into sensing a body. Such bodies can be artificial as well as real. For example, experimental work has shown subjects to experience the bodies of digital avatars, plastic mannequins or dolls (Slater & Usoh, 1993; Ehrsson et al., 2005); a concept described as *homuncular flexibility* to refer to high malleability of our body schema and the mental plasticity required to occupy virtual bodies (Won et al., 2015). Body swapping with another is made possible through a combination of VR first-person image avatars and visual motor agency in which the user's position and movements are captured through high-speed cameras to control the avatar and allowing them to interact through the body of another performer (De Oliveira et al., 2018). In TMTBA the user (person A) wears a head-mounted display that shows the perspective of a performer (person B) who mimics person A's movements. Person A's movements are monitored by a motion capture system for calibration. The head-tracking orientation of person A is sent to a servo motor on person B. Person B wears A, and these camera images are sent back (with no/low latency) to the head-mounted display of person A. This is accompanied by audio narratives and tactile stimulus to induce the body swapping illusion.

The De Oliveira et al. (2018) study of user experiences of VR technology, reveals it to be very effective in promoting this illusory response—in a system described as a form of *embodiment virtual reality*. TMTBA is not simply a technological tool for experiencing the novelty of body swapping; rather its creators are interested in prosocial value and behaviour change. Their aim is to lessen inter-group social barriers through the stimulation of an empathic response using the body ownership illusion. By imagining the body of others that might be otherwise marginalised within the hegemonic culture (whether due to race, ethnicity, disability, gender identity or sexuality) the aim of TMTBA is to foster empathy as a pathway towards altruistic behaviour and the overcoming of social difference. For example, in November 2016 at Cidade de Deus in Rio De Janeiro, a performance/experiment of The Machine to Be Another, allowed participants to vicariously experience the situation of parents of young black men killed by law enforcement officials, allowing the user to experience the interactions with law enforcement officers through the body of one of the parents of the victim.

TMTBA has the function of facilitating the 'veil of ignorance' (to borrow Rawls' term) to shape the user's understanding of the plight of others. As noted in previous chapters, the core challenges to an empathic ethics are the cognitive biases and heuristics that prevent us from empathising with out-group individuals. This is known as *implicit bias* (Greenwald & Krieger, 2006) against those with specific (often racially categorised) physical characteristics or ethnic identities. When combined with problems of *bounded rationality* such that one cannot process all the necessary complex information to make decisions (Forester, 1984; Kahneman, 2003), or the aforementioned *construal level problem*—such that issues and people that seem 'far away' geographically, socially, temporally or culturally are perceived in more abstract terms (Trope & Liberman, 2010), creating personal dissociation (and hence a lack of empathy towards them). In

Rawls' (1999) justice-as-fairness model we are charged with making decisions from an Original Position from which we cannot know the outcome for any specific social group, thus we would choose to act, not in self-interest, but for a collective sense of distributive justice between and amongst social groups. In reality, this is hindered by such cognitive biases and the failures that our collective bounded rationality create for morally just outcomes. TMTBA has the capacity to facilitate moral thinking to achieve just outcomes between social groups by increasing the field of experience and mental capacity of decision-makers to judge the experience of others, and thus make decisions informed by that experience. The value in TMTBA lies in the imaginative and stimulating element of empathic engagement with this decision-making process, which mirrors Deweyan concepts of dramatic rehearsal—improving the capacity of the individual to make moral decisions by enhancing the imaginative engagement with outcomes experienced by other people, which would otherwise be obscured by our moral habits (Dewey, 1917; Fesmire, 2003) and dictated by our implicit biases.

A Breath-Taking Journey

A Breath-taking Journey (ABTJ) is described as the amalgamation of a *persuasive game* and an *immersive technology*—an embodied and multisensory mixed-reality game providing a first-person perspective of a refugee's journey (Kors et al., 2016). Gamification is used in fields as diverse as marketing, healthcare and social activism as a means to increase user engagement in a saturated media landscape where passive advertising is no longer effective. Prosocial gaming, like its immersive journalism counterpart, focuses upon the experiences of marginalised populations such as migrants, disaster victims or individuals from Black, Asian and minority ethnic communities. Within the VR space, such prosocial media and gaming platforms aim to prompt the user to 'see what they see' and leverage immersive technology to 'sense what they sense'.

Showing the experience of suffering (e.g. the impact of natural disasters) or triumph over suffering (e.g. overcoming illness, building freshwater systems, post-disaster recovery) is a way to arouse empathy in the viewer/user. In both traditional and immersive journalism this has proven effective in either changing or reinforcing related attitudes and behaviours towards certain problems (e.g. famine, war, climate change) or the situation of social groups (e.g. cancer sufferers, refugees, endangered animals)

(Hoffman, 2008; Eisenberg & Fabes, 1990). A persuasive game heightens the empathic response when compared to legacy media (such as documentaries, adverts or televised appeals). The primary difference is in the presentation of a goal-oriented and interactive environment that uses role-play and role-taking rather than just visual representation to stimulate a response. The game structure encourages users to evaluate their choices in relation to a range of consequences and to "question the system a game represents" (Boltz et al., 2015). Games are particularly well-suited to fostering empathy because they allow players to inhabit the roles and perspectives of other people or groups in a uniquely immersive way (Belman & Flanagan, 2010).

ABTJ is what is termed a mixed-reality (Benford & Giannachi, 2011) rather than VR game because it incorporates simulated and physical stimuli together in one system. The narrative positions the user as a refugee fleeing from a (non-specific) war-torn country by hiding in the back of a lorry. The user wears a head-mounted display (as per other VR experiences) but also enclosed over-the-ear headphones to provide auditory immersion. This virtual system is then augmented with a range of physical elements, including a mask that incorporates a breathing sensor and scent diffuser, an enclosed physical space that creates a proxy for the inside of the lorry, a motor that simulates movement and throws the user off balance, and a controlled shutter that drops objects on the user at periodic intervals (Kors et al., 2016). As Kors et al.'s (2016) qualitative analysis of user experiences of ABTJ shows, the mixed-reality framework means that users place heavy emphasis upon bodily feelings and on physical sensations to create a repertoire of embodied feelings about the plight of refugees. As the authors note, the mixed-reality experience offers an embodied experience of 'inhabiting' another person's perspective by created a proxy sensation of what they experience on the refugee journey (and hence potentially feeling what they feel), thus sharing strong similarity with both Clouds Over Sidra and The Machine to Be Another. However, they also note when the user adopts another individual's point of view through an immersive device; it is difficult to ascertain with whom they are empathising—is it the person that they embody (ABTJ is a solitary experience) or is it a more generalised sense of 'the refugee' with which they empathise? We must question therefore in the application of such mixed (or solely virtual) reality technologies as ethics tools, the extent to which imaginative and empathic moral responses can be expanded or generalised beyond the specific individual user experiencing the virtual space and the pre-framed narrative embedded within the technology's script.

Discussion

VR has been applied by journalists, documentary film-makers, artists and philanthropists with the primary intention of stimulating empathy towards marginalised populations through a combination of immersion, embodiment and persuasion. VR can therefore be thought of as a moral enhancement technology—a system designed to change the behaviours of the user. Much of the existing literature on technological moral enhancement has focused upon pharmaceutical, neurological or genetic interventions to improve prosocial behaviour, though more recently moral enhancement through artificial intelligence is also discussed. However, VR is positioned here as a technology for moral enhancement principally by improving the empathy and moral imagination of the user. Whether or not the VR provides the basis for 'the ultimate empathy machine' requires further empirical investigation. VR technologies have gained momentum amongst NGOs, journalists, charities and other society organisations concerned with enhancing the welfare of marginalised people. Yet this enthusiasm is not backed by strong data on long-term empathy arousal—it is unclear at this stage whether or not the act of using a VR system for moral enhancement is sufficient to maintain longer-term behavioural change.

Empathy is partly physical; it has roots in a shared sense of both well-being and distress. Empathy is hard to imagine in the abstract—it is something that is felt between two people through the memory of shared experience, the capacity to imagine one another's feelings, and the sense of connection that emerges as a result. As described in the previous chapter, this feeling is subject to in-group/out-group bias, and the bias of individual moral responsibility to alleviate individual suffering over group suffering. Yet empathy also stimulates the conditions for prosocial behaviour, and it is an important precursor to ethical decision-making. From a Deweyan perspective an ethics VR must be not only immersive but must also stimulate a hunting phase for ethical problem-solving. As Rueda and Lara (2020) argue, boosting empathy through technological means is ethically desirable only if the technology appropriately identifies the social target, if the context in which empathy needs to be developed is justified, and if empathy is conceived as a necessary complement to other capabilities needed for doing the right thing. This is what is missing in all of the

VR platforms mentioned in this chapter. In each case, the moral subject is predefined by the journalist, artist, game designer or documentary filmmaker—each aims to predefine a set of ethical outcomes in advance, and then persuade the user of the VR to follow that line of thinking. This is problematic for number of reasons. First is the assumption that the creator has benign intentions. The politics 'scripted' into the VR system are implicit; there is a narrative framing that occurs. One could hypothetically create a VR system that promoted racist, sexist, homophobic or transphobic narratives. These would similarly stimulate immersive embodiment and empathy but would be morally undesirable. The second is what is sometimes referred to as the *deficit model* problem. Deficit models have been discussed in science education, public engagement and communication strategy—the term describes how powerful authorities generate specific narratives around a topic and then communicate these narratives in a linear fashion (i.e. providing a single line of evidence without opportunity for two-way exchange or dialogue to occur). The deficit model thinking assumes that once this narrative is heard and understood then the recipient will fundamentally agree with it. The assumption is that proponent and recipient will share the same values because they share the same information (see, e.g. Sturgis & Allum, 2004). It is a deficit model; it assumes that the problem is that the recipients' knowledge is faulty or incomplete and can thus be fixed through one-way communication. However, narratives, frames and stories are open to interpretive flexibility. The recipient interprets the narrative through the lens of pre-existing biases, cultural values and heuristics; such that no two individuals will 'receive' exactly the same message from any given communication source (Kahneman & Tversky, 1984; Hilligoss, 2014; McNally et al., 2018). The same is true of the virtual narratives of immersive journalism and gaming—there is no guarantee that universal behaviour change will occur as a result of exposure to a VR film or game. Overtly political narratives about hot button issues such as terrorism, poverty alleviation, abortion and so on, may have the desired effect to increase empathy, or they may stimulate a negative reaction amongst users whose political identity and ideology clashes with that of the creator. VR could reduce empathy amongst certain users.

When designing a VR ethical tool it must be as 'unbiased' as possible if it is to work as a trusted platform for decision-making. Immersive journalism operates within a context where trust in mainstream media is at an all-time low; the polarisation of public attitudes towards journalism in the age of 'alternative facts' and 'fake news' means that heterogeneous publics are increasingly sensitive to narrative framing, partisanship and the editorialisation of news stories. It is possible that immersive journalism may amplify rather than diminish this sensitivity. As Baudrillard (1995) discussed, the duplication made possible by media such as those presented by immersive witness, become a type of *hyperreality*—an immediate substitute for truth. The potential for psychological manipulation whereby immersive sensory input can overwhelm judgement rather than illuminate the moral imagination, is in itself is worthy of empirical investigation amongst journalism and ethics scholars. However, for the purposes of the development of an ethical tool, we can simply say that *neutrality* and pluralism are important prerequisites. One can assume neither that commonly shared moral judgements will emerge through empathy alone when exposed to an immersive VR experience, nor that the reactions of the user can be planned by the VR designer. The broader context of the VR platform development is a therefore a critical element of an ethical tool—it must be shown to have diversity and plurality of ethical values, provide autonomous choice within the virtual environment and be transparent in its structure and objectives from the outset. An ethical tool is only useful if it can be trusted to be transparent in its aims and provide adequate user autonomy.

References

Aitken, S., & Craine, J. (2009). Into the image and beyond: Affective visual geographies and GIScience. In M. Cope & S. Elwood (Eds.), *Qualitative GIS: A mixed methods approach* (pp. 139–155). SAGE.

Anderson, M. (2015, December 31). Can tearjerker virtual reality movies tempt donors to give more aid?, *The Guardian*. https://www.theguardian.com/global-development/2015/dec/31/virtual-reality-movies-aid-humanitarian-assistance-united-nations

Bailey, J., Bailenson, J. N., Stevenson Won, A., Flora, J., & Armel C. K. (2012). Presence and memory: Immersive virtual reality effects on cued recall. *Proceedings of the International Society for Presence Research Annual Conference* (pp. 24–26). Philadelphia, PA.

Baudrillard, J. (1995). *Simulacra and simulation—The body, in theory: Histories of cultural materialism*. University of Michigan Press.

BeAnotherLab. (2017). *The machine to be another*. http://www.themachinetobeanother.org/?page_id=764

Belman, J., & Flanagan, M. (2010). Designing games to foster empathy. *International Journal of Cognitive Technology, 15*(1), 5–15.

Benford, S., & Giannachi, G. (2011). *Performing mixed reality*. The MIT Press.
Benítez de Gracia, M. J., Herrera Damas, S., & Benítez de Gracia, E. (2019). Analysis of the immersive social content feature in the Spanish news media. *Revista Latina de Comunicación Social, 74*, 1655–1679.
Boltz, L. O., Henriksen, D., Mishra, P., & Group, D.-P. R. (2015). Rethinking technology & creativity in the 21st century: Empathy through gaming-perspective taking in a complex world. *TechTrends, 59*(6), 3–8.
Daft, R. L., & Lengel, R. H. (1986). Organizational information requirements, media richness and structural design. *Management Science, 32*(5), 554–571.
De la Peña, N., Weil, P., Llobera, J., Giannopoulos, E., Pomés, A., Spanlang, B., Friedman, D., Sanchez-Vives, M. V., & Slater, M. (2010). Immersive journalism: Immersive virtual reality for the first-person experience of news. *Presence: Teleoperators and Virtual Environments, 19*(4), 291–301.
de Oliveira Musse, Jorge, Aline Sacchi Homrich, Renato de Mello, & Marly M Carvalho. (2018). "Applying backcasting and system dynamics towards sustainable development: The housing planning case for low-income citizens in Brazil." *Journal of Cleaner Production, 193*, 97–114.
Dewey, J. (1917). *Creative intelligence*. Henry Holt and Co.
Ehrsson, H. H., Holmes, N. P., & Passingham, R. E. (2005). Touching a rubber hand: Feeling of body ownership is associated with activity in multisensory brain areas. *Journal of Neuroscience, 25*(45), 10564–10573.
Eisenberg, N., & Fabes, R. A. (1990). Empathy: Conceptualization, measurement, and relation to prosocial behavior. *Motivation and Emotion, 14*(2), 131–149.
Fesmire, S. (2003). *John Dewey and moral imagination: Pragmatism in ethics*. Indiana University Press.
Forester, J. (1984). Bounded rationality and the politics of muddling through. *Public Administration Review, 44*(1), 23–31.
Förster, J., Friedman, R. S., & Liberman, N. (2004). Temporal construal effects on abstract and concrete thinking: Consequences for insight and creative cognition. *Journal of Personality and Social Psychology, 87*(2), 177.
Greenwald, A. G., & Krieger, L. H. (2006). Implicit bias: Scientific foundations. *California Law Review, 94*(4), 945–967.
Hilligoss, B. (2014). Selling patients and other metaphors: A discourse analysis of the interpretive frames that shape emergency department admission handoffs. *Social Science & Medicine, 102*, 119–128.
Hodgkinson, G. (2016). Lock up your stories-here comes virtual reality. *TECHART: Journal of Arts and Imaging Science, 3*(4), 10–14.
Hoffman, M. L. (2008). Empathy and prosocial behaviour. In M. Lewis, J. M. Haviland-Jones, & L. Feldman-Barrett (Eds.), *Handbook of emotions* (pp. 440–455). Guildford Press.

Jones, S. (2017). Disrupting the narrative: Immersive journalism in virtual reality. *Journal of Media Practice, 18*(2–3), 171–185.

Kahneman, D. (2003). A perspective on judgment and choice: Mapping bounded rationality. *American Psychologist, 58*(9), 697–720.

Kahneman, D., & Tversky, A. (1984). Choices, values, and frames. *American Psychologist, 39*(4), 341.

Kool, H. (2016). The ethics of immersive journalism: A rhetorical analysis of news storytelling with virtual reality technology. *Intersect: The Stanford Journal of Science, Technology, and Society, 9*(3). http://ojs.stanford.edu/ojs/index.php/intersect/article/view/871/863

Kors, M. J., Ferri, G., Van Der Spek, E. D., Ketel, C., & Schouten, B. A. (2016). A breathtaking journey. On the design of an empathy-arousing mixed-reality game. In Proceedings of the 2016 Annual Symposium on Computer-Human Interaction in Play Oct 15th, 2016 (pp. 91–104).

McNally, H., Howley, P., & Cotton, M. (2018). Public perceptions of shale gas in the UK: Framing effects and decision heuristics. *Energy, Ecology and Environment, 3*(6), 305–316.

Milk, C. (2015). How Virtual Reality can create the Ultimate Empathy Machine. *TED Talk*. https://www.ted.com/talks/chris_milk_how_virtual_reality_can_create_the_ultimate_empathy_machine, Accessed 20/06/2020.

Nash, K. (2018). Virtual reality witness: Exploring the ethics of mediated presence. *Studies in Documentary Film, 12*(2), 119–131.

Rawls, J. (1999). *A theory of justice* (2nd ed.). Oxford University Press.

Rueda, J., & Lara, F. (2020). Virtual reality and empathy enhancement: Ethical aspects. *Frontiers in Robotics and AI*, 7. https://doi.org/10.3389/frobt.2020.506984

Shin, D.-H. (2017). The role of affordance in the experience of virtual reality learning: Technological and affective affordances in virtual reality. *Telematics and Informatics, 34*(8), 1826–1836.

Sirkkunen, E., & Uskali, T. (2019). Virtual reality journalism. In *The international encyclopedia of journalism studies*. Wiley Blackwell.

Slater, M., & Usoh, M. (1993). Representations systems, perceptual position, and presence in immersive virtual environments. *Presence: Teleoperators & Virtual Environments, 2*(3), 221–233.

Steuer, J. (1994). Defining virtual reality: Dimensions determining telepresence. *Journal of Communication, 42*(4), 73–93.

Sturgis, P., & Allum, N. (2004). Science in society: Re-evaluating the deficit model of public attitudes. *Public Understanding of Science, 13*(1), 55–74.

The World Economic Forum. (2020). *Our mission*. Geneva. Retrieved June 12, 2020, from https://www.weforum.org/about/world-economic-forum/

Trope, Y., & Liberman, N. (2010). Construal-level theory of psychological distance. *Psychological Review, 117*(2), 440.

UNVR. (2020). *UN virtual reality*. New York: United Nations Virtual Reality (UNVR), A Project Implemented by the UN SDG Action Campaign. Retrieved June 12, 2020, from http://unvr.sdgactioncampaign.org/vr-films/#.YATy5y-l2cY

Won, A. S., Bailenson, J., & Lanier, J. (2015). Homuncular flexibility: The human ability to inhabit nonhuman avatars. In R. A. Scott, S. M. Kosslyn, & M. Buchmann (Eds.), *Emerging trends in the social and behavioral science: An interdisciplinary, searchable, and linkable resource* (pp. 1–16). John Wiley & Sons.

CHAPTER 6

Developing a Virtual Reality Ethical Tool

Abstract This chapter outlines the structure for a virtual reality (VR) ethical tool grounded in John Dewey's concept of dramatic rehearsal. The culmination of ethical thinking on empathy, moral imagination and principlism is brought together with practical insights gained from the analysis of real-world examples of empathy-arousal virtual reality systems discussed in Chap. 5. The steps and sequence for a VR ethical tool are presented, and application to real-world practice is discussed through a vignette on medical ethics training on the issue of informed patient consent.

Keywords Dramatic rehearsal • Ethical tools • Reflective equilibrium • Informed consent

Introduction

Each of the VR platforms discussed in Chap. 5 (including the mixed-reality system ABTJ) are *empathy-arousal* tools, intended to promote pro-social behaviours and practices amongst users by stimulating empathic engagement towards marginalised or vulnerable peoples, including those at the heart of political controversy or victims of state violence. The assumption driving prosocial VR platforms is that the experience will

M. Cotton, *Virtual Reality, Empathy and Ethics*,
https://doi.org/10.1007/978-3-030-72907-3_6

stimulate a specific response amongst users—one that facilitates a personal sense of empathy towards the subject. Empathy-arousal using immersive witness is as Harper (2002, p. 13) describes, based upon a mental model that has "a physical basis: the parts of the brain that process visual information are evolutionarily older than the parts that process verbal information." The combination of sensory stimulation and narrative-style elicitation allows new opportunities for respondents to express and reflect upon ethical values, articulating a broad range of non-cost and non-utilitarian values in the process (Satterfield, 2001).

Empathy arousal through sensory stimulation is achieved in different ways in each of the cases discussed in the previous chapter. In the UNVR Series and the work of Nonny De La Peña and associates, the immersive journalistic reporting style emphasises the physical and social conditions of groups or individuals in distress by reporting the lived experience within specific scenarios and spaces—refugee camps, prisons, food banks and health centres. In The Machine To Be Another, and A Breath-taking Journey, there is a strong sense of embodiment within an avatar through a combination of haptic and (in the latter case, olfactory) sensory input overlaid on the visual stimulus of the virtual environment. It is through these mechanisms that interaction becomes even more personal. The relationship between the physiological reaction to immersive sensory environment and the capacity to imagine the consequences of different scenarios is deeply beneficial to ethical decision-making under a Deweyan model. For example, the body swapping perception encourages a deeply imaginative response to issues of racial justice, disrupting our mental habits and encouraging personal reflection upon one's implicit biases.

In the Deweyan model of dramatic rehearsal, the aim is to create a hunting phase of moral reflection. VR as ethical tool can enable users to gather morally relevant information towards a specific outcome or decision. However, as discussed in the previous chapter, framing the narrative in such a way as to arouse an empathic response from users may lead them to feel manipulated, especially when they are immersed in powerful auditory, visual and haptic sensory inputs. A VR ethical tool must reduce this risk by diversifying the narrative frames deployed such that (a) the morally relevant aspects of the narrative are at the forefront of the user's mind, and (b) that value pluralism and choice are at the heart of the user interaction. This can be achieved by presenting a range of different scenarios, locations, sensory inputs, and encountered individuals (or virtual non-human characters), so that user experiences many different perspectives and values. This aids the 'hunting process' of problem definition.

In many participatory ethics assessment processes it is common to do some kind of 'stakeholder mapping'. Those with decision-making power and those affected by decisions are conceptually grouped in order to understand the different roles and responsibilities of different moral agents involved in any given scenario (Beekman & Brom, 2007; Cotton, 2013; Mepham, 1999). What VR does, through processes of embodiment, is allow the user to better imagine themselves in the role of those different stakeholders, or else to encounter them as 'characters' within the narrative structure. It is through this process of encountering others that VR facilitates moral imagination. Users can more easily identify different stakeholder positions and then virtually 'live out' the experience of each; or else seeing first-hand the consequences of autonomous choices made within the scenario.

Autonomy and choice are pivotal. In Clouds Over Sidra (COS) the emphasis is upon presence within a journalistic setting—to reduce the distance between the reporter, the geographic situation, the individuals within the refugee camp, and the viewer living far away from the situation. But the choice of where to go and what to do is limited—certainly when compared to gaming environments that employ VR through open world or 'sandbox' approaches. In A Breath-taking Journey (ABTJ) the design goes some way towards gamifying the experience, in the sense of adding choices and branching narrative. The gamification of empathy-arousal VR environments would encourage greater visualisation, experimentation and creativity of play within the environment. This increases the imaginative engagement with the system by encouraging a greater level of problem solving and by 'seeing' causal relationships between individual actions and whole systems (Betz, 1995). Just as greater heterogeneity of stakeholder positions is necessary to understand different roles and responsibilities for moral agents, so the increase in choice of outcomes encourages a deeper understanding of the consequences of moral agents' actions.

Computer role-playing games (RPGs), discussed in previous chapters, are adept at providing choice and narrative complexity in a coherent manner. Just as branching dialogue options can lead to diverse outcomes in a game environment, so must an ethical tool allow different choices and consequences to be displayed. Whereas the narrative of many VR documentaries, films and games are either linear or involves a narrow degree of interactive choice, an ethical tool must allow choices to be wide ranging and comprehensive, reflective of an array of ethical content and coherent in the application and approach. Models based upon principlism (Beauchamp & Childress, 2001), or reflective equilibrium (Cotton, 2009a; Daniels, 1996)—such as the Ethical Grid (Seedhouse, 1998; Machlaren and Seedhouse, 2001), Ethical Matrix (Mepham, 1999; Schroeder & Palmer, 2003; Cotton, 2009b), Ethical Delphi (Millar et al., 2007) or Reflective Ethical Mapping (Cotton, 2014) have potential value due to their *value pluralism*—presenting a range of interactions between competing principles, judgements and stakeholder values. What VR provides is the opportunity to interact through empathic and imaginative engagement with the values in a way that pen-and-paper or grid-like models do not. The principles presented in these tools—such as justice (fair outcomes for different parties), autonomy (freedom of choice through informed consent), well-being (the meeting of basic needs), beneficence (a desire to improve the welfare of others), non-maleficence (first do no harm), fidelity (keep promises) or honesty (to be truthful)—provide useful guides for the development of dialogue choices within the narrative. In essence, different scripted lines of interaction within the VR and dialogue choice options can be structured around this principlist approach—the option to emphasise the specific principle will lead to different outcome for the subject of the ethical decision. By then using a 'cut scene' approach that shows the consequences of the ethical choice, this allows imaginative engagement with the future—one can 'try on' within the virtual environment and then reflect upon how it resonates with one's own judgements, personal values and feelings. It is in this way that the virtual reality becomes an ethical tool—it is a judgement aid that encourages reflection about the consequences of choice within an immersive, imaginative, safe and replicable environment; it is not a decision-tree that leads to a prescriptive course of action.

The practical design of a dramatic rehearsal VR ethics tool would necessarily involve the input of a range of different stakeholders. The design process is itself a hunting phase. The input of professional associations, oversight bodies, citizens groups, NGOs and other interested parties is needed to get a sense of the ethical issues at stake, the problems to resolve and the types of interactions that would occur in the course of making an ethical decision. Of paramount importance is the quality of the dialogue—it must blend together insight from the principlist approach that emphasises specific aspects of common-sense moral judgement, with naturalistic dialogue such that the reactions of the encountered 'characters' within the

VR ring true for real-world practice. This approach has precedence in current ethics training in a range of different fields. For example, medical professionals have training in role-play simulation of doctor-patient encounters, such as delivering bad news, dealing with patient complaints and gaining informed consent (Buxton et al., 2015), which provide a useful training resource for when these situations are encountered in the real world. In ethics teaching and formalised ethical decisions-making (particularly within medical, engineering or business ethics) it is common to use 'live' scenarios. The trainee selects amongst a range of cases/scenarios that involve ethical violations, formulates a position and then debates the choices made. This process of reflection increases broader awareness of the complexity of ethical challenges, allows the application of concepts and creates a personal, emotional and critical engagement with the ethical issues at hand (McWilliams & Nahavandi, 2006). A VR programme based upon dramatic rehearsal would allow a standardised, replicable, low-cost and shareable approach to this type of scenario-based simulation.

Figure 6.1 outlines the design structure for VR programme development that integrates a principlist approach with dramatic rehearsal. The platform structure is broken down into four stages. The first is a development stage. The designer must first define the domain of ethics and then create a virtual world that captures the nature of that domain. The input of domain-specific ethicists (medical ethics/bioethics, environmental ethics, business ethics etc.) to define the relevant principles and theoretical constraints, and professional stakeholders that can help to define the core practical considerations is necessary at this early stage. It is important that the design captures not only the principles at stake, but the factors that would make the scenario realistic, including all of the contextual elements that would allow naturalistic interaction within the virtual environment. The second part is the design stage. This turns the combination of ethical principles and practical experience into a series of scenarios—interactions and encounters structured by a branching narrative approach. A virtual environment in which the encounters between the user and the virtual characters take place must be built (for example, within a healthcare professional's office, or on a ward, at a hearing or tribunal and so on). This virtual environment should be complete with familiar sounds, objects and characters so that the engagement with the ethics of the situation is grounded in contextual realism. This will enhance empathic engagement through sensory immersion. Realistic characters must be designed based upon real-world experience and with an extensive dialogue tree, with

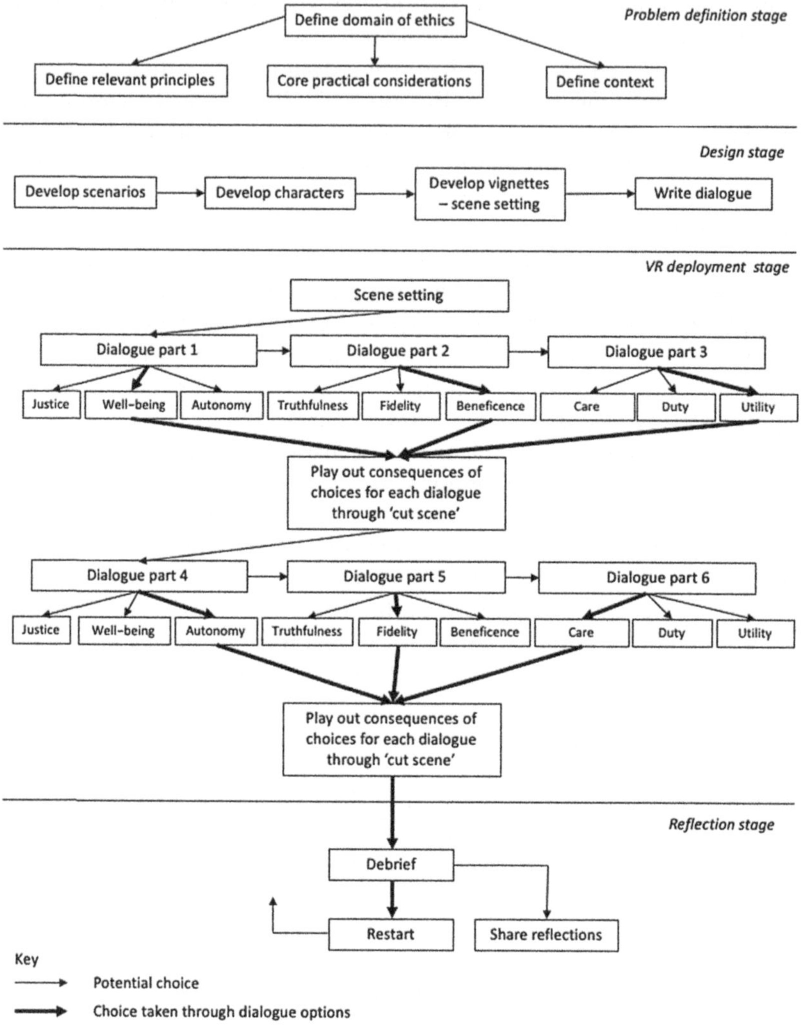

Fig. 6.1 Dramatic rehearsal VR programme structure

branching options for the user to choose different ways of interacting. These branching dialogue options will then lead to 'cut scenes' that play out of the consequences of the choices made.

The third stage of the dramatic rehearsal VR programme is deployment. Once designed, the VR becomes an ethical tool only through use by the practitioner. The programme begins with the scene-setting in which the characters and environment are introduced to orient the user. It is at that point that the user engages in dialogue with virtual characters. The VR might allow the user to adopt the position of the ethical decision-maker, or the subject of the ethical decisions—for example either as a physician or as a patient. It is at this point that the user is given a set of branching dialogue options with each emphasising a specific moral principle in natural language. In Fig. 6.1 some examples of principles are shown, and these must be relevant to the cases, context and dialogue. By moving through the branching options and witnessing direct feedback from the virtual characters, the user can get a feel for the real-life interaction of ethics and social engagement in managing specific problems. Once the user reaches the end of the branching dialogue there then shown a cut scene that plays out the future that is created by the ethical choices taken by the user. They would then have the option either to restart the scenario from the perspective of a different character or to choose different options within the branching dialogue. Once the user is satisfied that a range of different ethical outcomes have been explored this then leads to the fourth and final reflection stage—in which the user is debriefed about the consequences of the choices made. This stage could be within the VR platform itself (guided by a virtual character), or through the guidance and advice of a peer or mentor in an offline environment. It is by sharing the reflections from 'trying on' the ethical futures created by this virtual dramatic rehearsal process, that the tool has value. It is the combination of the hunting phase to define the ethical problems through choice architecture, reflection upon the consequences of those choices in relation to personal moral judgements and core values, and the implementation of those choices in practice that the dramatic rehearsal is complete.

Dramatic Rehearsal in Practice: The Case of Medical Ethics Training

The dramatic rehearsal model is applied here in a brief vignette relevant to medical ethics—using the issue of ensuring medical consent for patients as a scenario for the VR platform. To give some context: consent from a patient signals informed agreement for a treatment or intervention to go ahead. The requirement to seek patient consent is therefore a strong expression of the moral obligation to respect patient autonomy and is universally required by law. Many ethical challenges faced by healthcare professionals arise where there may be doubts about an individual's ability to consent to treatment (including for example very young children, those with specific learning difficulties, or those suffering from psychological or physiological impairments to communication such as those who are unconscious or anaesthetised and in need of treatment). The individual giving consent has to have the cognitive capacity to be able to make the decision in question. Consent has to be given voluntarily and without coercion. Sufficient information also has to be offered to enable the individual to understand the nature of the decision and its likely consequences, including the consequences of declining the treatment or intervention. However, some circumstances (including certain emergencies) make it difficult to get consent. Consent can be explicit, given either orally or in writing, or implied, where it may be signalled by the behaviour of an informed patient, but this can be difficult to interpret due to language barriers, cultural differences of understanding, or a range of other circumstances. Consent must be sought for a range of different processes and outcomes—including if a student is present during a consultation, if a patient is anaesthetised, or if a certain procedure contradicts the beliefs of an individual (such as a critically ill patient who is a Jehovah's Witness refusing a blood transfusion) (Singer, 2000; BMA, 2019). Consent is an ongoing process, not a single one-off decision, for example if there is significant interval between a patient agreeing to a treatment and its start, or if new information has emerged, or the patient's condition has altered, consent should be reaffirmed. Consent can be denied even when doing so will result in the patient's permanent injury or death. "The problem for all who care about others just is how to reconcile respect for the free choices of others with real concern for their welfare, when their choices are or appear to be self-destructive." (Harris, 2006).

Navigating the complexity of consent scenarios requires dialogue and an understanding of how different choices will lead to diverse outcomes. When exploring the consent issue through a dramatic rehearsal VR, it would be important to first begin by talking with patient advocacy organisations, legal professionals in medical practice, healthcare professionals and oversight bodies/board certification bodies, to map out what the issues of consent are, to understand the legal ramifications of non-compliance, to present a series of real-world scenarios from practical experience. It would be useful to integrate qualitative data/recordings of real healthcare professional-patient interactions to provide the naturalistic dialogue necessary for stimulating an immersive empathic response by users. The design of the VR might take place in a healthcare professional's place of work, though may also switch to other settings such as the classroom, court room or tribunal—as these are relevant settings to experience some of the outcomes of choices made. The development of dialogue would involve interaction with different types of patient circumstances, covering situations in which consent may be difficult to obtain. These scenarios might include a child that one may suspect is subject to abuse or neglect; a patient with dementia undergoing end of life care; or a family making decisions about a loved one in an unresponsive/unconscious state. Through the process of interaction, the user can choose dialogue branches that relates to specific principles. So for example, one dialogue branch may emphasise the principle of *autonomy*, in which the user provides the maximum amount of information necessary to the patient or patient's advocate, but provides no indication of personal preference for treatment options; a second branch of dialogue may emphasise *well-being*, in which the user presents treatment options as a balance of risks and advises the patient to opt for the treatment option with the lowest risk to reward ratio; and the third might emphasise *utility* in which a treatment which provides the best outcome for the lowest cost is presented as the preferred option. In each of the dialogue branches, the VR provides immediate feedback from the patient's reaction including negative emotional reactions such as anger or sorrow, creating a connection between the user and the hypothetical healthcare professional-patient interaction. This creates an embodied ethical scenario that is context-specific and relevant to practice. By using a cut scene to show the consequence of each choice, the hunting phase of the dramatic rehearsal is complete. For example, if the user chose the autonomy dialogue option, the patient may be overwhelmed by the information presented and may choose to not have any treatment due to concern over the side effects or adverse consequences of doing so. If the user chose the well-being dialogue option, then patients may desire high-cost state-of-the-art treatments which may have limited availability and then become angry or frustrated if those options are not available locally. Similarly, if the user chose the utility option the outcome to the patient maybe adverse health consequences which would not have been experienced if either of the other two options were selected.

Not all ethical dilemmas present an option that clearly overrides all others. A pluralistic ethics model, such as that presented by dramatic rehearsal, does not hierarchically rank principles but rather uses them as a guide to facilitate moral imagination by relating choice to consequence in a hypothetical scenario, and then reflecting upon the outcomes of that choice. This process of reflection is what Rawls terms *reflective equilibrium* (Rawls, 1951; Daniels, 1979), in which principles and personal judgements are brought into harmony with one another through an internal deliberation about future consequences of specific choices. The VR primes the user to make informed moral decisions through this imaginative and immersive space, but this is not a substitute for reflection and reasoned judgement. As with any tool, it facilitates but does not replace the moral faculties of the individual decision-maker.

Conclusions

A VR programme designed around dramatic rehearsal approach must involve a combination of 'hunting', empathic engagement, a choice of principles, the imagination of outcomes and reflection upon practice, if it is to be successful. VR allows users to imagine the consequences emerging from different courses of action, accompanied by a sense of embodiment which, in turn, increases the user's empathic engagement with the situation overall. Crucially, the value of the VR as an ethical tool is not isolated to the user experience in the virtual environment. To use VR in teaching, training or professional environment requires offline deliberation in order to provide meaning and context to the decisions taken within the virtual environment. This requires support. The deliberation of choice and consequence with peers, teachers and mentors is necessary in order to consolidate the process of social learning that goes on within the virtual environment, and to ground an understanding about the consequences of ethical choices in the experience of others. This 'debriefing' also provides an opportunity to reflect upon the embodied experience of the difference stakeholders captured within the virtual environment. It is in this way that the Deweyan concept of dramatic rehearsal harmonises with Rawls' concept of *reflective equilibrium;* the aim of the reflective process is to bring balance and coherence within the user's beliefs on general principles that arrive through a process of deliberative mutual adjustment. The VR is a tool to achieve this, in the sense that it is a resource that facilitates the hunting of alternatives, imagination of different experiences, empathy towards other social groups, and the dramatic rehearsal of different outcomes that stem from moral choices. However, the tool alone does not do the work; it is a heuristic device that aids the actions of moral agents, rather than substituting moral judgement and intuition. The VR is a judgement aid because it compensates for the limitations of our thinking; it disrupts our habits of thought, implicit biases, and improves our imaginative capacity to engage with the experiences of others. I propose, therefore, that a programme of empirical research into participatory user design of VR as an ethical tool in a range of professional contexts is necessary, as well as qualitative evaluative research on the impact that such a tool would make to the moral judgements of the user.

References

Beauchamp, T. L., & Childress, J. F. (2001). *Principles of biomedical ethics* (5th ed.). Oxford University Press.

Beekman, V., & Brom, F. W. A. (2007). Ethical tools to support systematic public deliberations about the ethical aspects of agricultural biotechnologies. *Journal of Agricultural and Environmental Ethics, 20*(1), 3–12.

Betz, J. A. (1995). Computer games: Increase learning in an interactive multidisciplinary environment. *Journal of Educational Technology Systems, 24*(2), 195–205.

BMA. (2019). *Medical students ethics tool kit*: British Medical Association. Retrieved July 19, 2019, from https://www.bma.org.uk/advice/employment/ethics/medical-students-ethics-toolkit

Buxton, M., Phillippi, J. C., & Collins, M. R. (2015). Simulation: A new approach to teaching ethics. *Journal of Midwifery & Women's Health, 60*(1), 70–74.

Cotton, M. (2009a). Ethical assessment in radioactive waste management: A proposed reflective equilibrium-based deliberative approach. *Journal of Risk Research, 12*(5), 603–618.

Cotton, M. (2009b). Evaluating the 'ethical matrix' as a radioactive waste management deliberative decision-support tool. *Environmental Values, 18*(2), 153–176.

Cotton, M. (2013). Deliberating intergenerational environmental equity: A pragmatic, future studies approach. *Environmental Values, 22*(3), 317–337.
Cotton, M. (2014). *Ethics and technology assessment: A participatory approach.* Springer-Verlag.
Daniels, N. (1979). Wide reflective equilibrium and theory acceptance in ethics. *Journal of Philosophy, 76*(5), 256–282.
Daniels, N. (1996). *Justice and justification: Reflective equilibrium in theory and practice.* Cambridge University Press.
Harper, D. (2002). Talking about pictures: A case for photo elicitation. *Visual Studies, 17*(1), 13–26.
Harris, J. (2006). *The value of life: An introduction to medical ethics.* Routledge.
Machlaren, P., & Seedhouse, D. (2001). Computer Mediated Communication with Integrated Graphical Tools Used For Health Care Decision-Making. In Short Paper Proceedings of the 18th Annual Conference of the Australasian Society for Computers in Learning in Tertiary Education (ASCILITE), 109–112.
McWilliams, V., & Nahavandi, A. (2006). Using live cases to teach ethics. *Journal of Business Ethics, 67*(4), 421–433.
Mepham, B. (1999). A framework for the ethical analysis of novel foods: The ethical matrix. *Journal of Agricultural and Environmental Ethics, 12*, 165–176.
Millar, K., Thorstensen, E., Tomkins, S., Mepham, B., & Kaiser, M. (2007). Developing the ethical Delphi. *Journal of Agricultural and Environmental Ethics, 20*(1), 53–63.
Rawls, J. (1951). Outline of a decision procedure for ethics. *The Philosophical Review, 60*(2), 177–197.
Satterfield, T. (2001). In search of value literacy: Suggestions for the elicitation of environmental values. *Environmental Values, 10*(3), 331–359.
Schroeder, D., & Palmer, C. (2003). Technology assessment and the 'ethical matrix'. *Poiesis & Praxis, 1*(4), 295–307.
Seedhouse, D. (1998). *Ethics: The heart of health care.* Wiley.
Singer, P. A. (2000). Medical ethics. *British Medical Journal, 321*(7256), 282–285.

References

Aitken, S., & Craine, J. (2009). Into the image and beyond: Affective visual geographies and GIScience. In M. Cope & S. Elwood (Eds.), *Qualitative GIS: A mixed methods approach* (pp. 139–155). SAGE.

Alcorn, P. A. (2001). *Practical ethics for a technological world*. Prentice Hall.

Almond, B. (1988). Women's right: Reflections on ethics and gender. In M. Griffiths & M. Whitford (Eds.), *Feminist perspectives in philosophy*. Indiana University Press.

Anderson, M. (2015, December 31). Can tearjerker virtual reality movies tempt donors to give more aid?, *The Guardian*. https://www.theguardian.com/global-development/2015/dec/31/virtual-reality-movies-aid-humanitarian-assistance-united-nations

Anderson, E. (2018). Dewey's moral philosophy. In *Stanford encyclopedia of philosophy* [Online]. Version. https://plato.stanford.edu/entries/dewey-moral/

Anderson, B., & Tracey, K. (2001). Digital living: The impact (or otherwise) of the Internet on everyday life. *American Behavioral Scientist, 45*(3), 456–475.

Anderson, C. A., Shibuya, A., Ihori, N., Swing, E. L., Bushman, B., Sakamoto, A., Rothstein, H. R., & Saleem, M. (2010). Violent video game effects on aggression, empathy, and prosocial behavior in Eastern and Western countries: A meta-analytic review. *Psychological Bulletin, 136*(2), 151–173.

Aristotle. (2000). *Nicomachean ethics* (R. Crisp, Trans.). Cambridge University Press.

Aromaa, S., & Väänänen, K. (2016). Suitability of virtual prototypes to support human factors/ergonomics evaluation during the design. *Applied Ergonomics, 56*, 11–18.

M. Cotton, *Virtual Reality, Empathy and Ethics*,
https://doi.org/10.1007/978-3-030-72907-3

Arriaga, P., Esteves, F., Carneiro, P., & Monteiro, M. B. (2006). Violent computer games and their effects on state hostility and physiological arousal. *Aggressive Behavior: Official Journal of the International Society for Research on Aggression, 32*(2), 146–158.

Augustine, D. L. (2018). *Taking on technocracy: Nuclear power in Germany, 1945 to the present.* Berghahn Books.

Baggini, J., & Fosl, P. S. (2007). *The ethics toolkit: A compendium of ethical concepts and methods.* Wiley-Blackwell.

Bailey, J., Bailenson, J. N., Won, A. S., Flora, J., & Armel, K. C. (n.d.). Presence and memory: Immersive virtual reality effects on cued recall. *Proceedings of the International Society for Presence Research Annual Conference*: Citeseer, 24–26.

Baird, C. A. (2005). *Everyday ethics: Making hard choices in a complex world.* Tendril Press.

Baron-Cohen, S. (2012). *The science of evil: On empathy and the origins of cruelty.* Basic Books.

Bates, J. (1992). Virtual reality, art, and entertainment. *Presence: Teleoperators & Virtual Environments, 1*(1), 133–138.

Battaly, H. D. (2011). Is empathy a virtue? In A. Coplan & P. Goldie (Eds.), *Empathy: Philosophical and psychological perspectives* (pp. 277–301). Oxford University Press.

Baudrillard, J. (1995). *Simulacra and simulation—The body, in theory: Histories of cultural materialism.* University of Michigan Press.

Baudrillard, J. (2005). Violence of the virtual and integral reality. *International Journal of Baudrillard Studies, 2*(2), 1–16.

BeAnotherLab. (2017). *The machine to be another.* http://www.themachinetobeanother.org/?page_id=764

Beardon, C. (1992). The ethics of virtual reality. *Intelligent Tutoring Media, 3*(1), 23–28.

Beauchamp, T. L. (1984). On eliminating the distinction between applied ethics and ethical theory. *The Monist, 67*, 514–531.

Beauchamp, T. L., & Childress, J. F. (2001). *Principles of biomedical ethics* (5th ed.). Oxford University Press.

Beekman, V., & Brom, F. W. A. (2007). Ethical tools to support systematic public deliberations about the ethical aspects of agricultural biotechnologies. *Journal of Agricultural and Environmental Ethics, 20*(1), 3–12.

Beierle, T. C. (1999). Using social goals to evaluate public participation in environmental decisions. *Policy Studies Journal, 3*(4), 75–103.

Beierle, T. J. (2002). The quality of stakeholder-based decisions. *Risk Analysis, 22*(4), 739–748.

Belman, J., & Flanagan, M. (2010). Designing games to foster empathy. *International Journal of Cognitive Technology, 15*(1), 5–15.

Benedict, R. (1999). A defense of ethical relativism. In H. J. Curzer (Ed.), *Ethical theory and moral problems*. Wadsworth Publishing Company.

Benford, S., & Giannachi, G. (2011). *Performing mixed reality*. The MIT Press.

Benítez de Gracia, M. J., Herrera Damas, S., & Benítez de Gracia, E. (2019). Analysis of the immersive social content feature in the Spanish news media. *Revista Latina de Comunicación Social, 74*, 1655–1679.

Bernstein, R. J. (1992). The resurgence of pragmatism. *Social Research, 59*, 813–840.

Betz, J. A. (1995). Computer games: Increase learning in an interactive multidisciplinary environment. *Journal of Educational Technology Systems, 24*(2), 195–205.

Bhagat, K. K., Liou, W.-K., & Chang, C.-Y. (2016). A cost-effective interactive 3D virtual reality system applied to military live firing training. *Virtual Reality, 20*(2), 127–140.

Biocca, F., & Levy, M. R. (2013). *Communication in the age of virtual reality*. Routledge.

Blackall, M. (2020). Channel 4 under fire for deepfake Queen's Christmas message *The Guardian*, Available: Guardian Media Group. https://www.theguardian.com/technology/2020/dec/24/channel-4-under-fire-for-deepfake-queen-christmas-message

Bloom, P. (2017a). *Against empathy: The case for rational compassion*. Random House.

Bloom, P. (2017b). Empathy and its discontents. *Trends in Cognitive Sciences, 21*(1), 24–31.

Blowers, A., & Sundqvist, G. (2010). Radioactive waste management—Technocratic dominance in an age of participation. *Journal of Integrative Environmental Sciences, 7*(3), 149–155.

BMA. (2019). *Medical students ethics tool kit*: British Medical Association. Retrieved July 19, 2019, from https://www.bma.org.uk/advice/employment/ethics/medical-students-ethics-toolkit

Boltz, L. O., Henriksen, D., Mishra, P., & Group, D.-P. R. (2015). Rethinking technology & creativity in the 21st century: Empathy through gaming-perspective taking in a complex world. *TechTrends, 59*(6), 3–8.

Botella, C., Serrano, B., Baños, R. M., & Garcia-Palacios, A. (2015). Virtual reality exposure-based therapy for the treatment of post-traumatic stress disorder: A review of its efficacy, the adequacy of the treatment protocol, and its acceptability. *Neuropsychiatric Disease and Treatment, 11*, 2533–2545.

Bowen, S. A. (2005). A practical model for ethical decision making in issues management and public relations. *Journal of Public Relations Research, 17*(3), 191–216.

Brandt, R. B. (1959). *Ethical theory: The problems of normative and critical ethics*. Prentice-Hall.

Brennan, S. (1999). A survey of recent work in feminist ethics. *Ethics, 109*(4), 858–893.
Brennan, S. (Ed.). (2002). *Feminist moral philosophy*. University of Calgary Press.
Brewster, D. (1856). *The stereoscope; Its history, Theory and construction, with its application to the fine and useful arts and to education. Etc.* John Murray.
Brey, P. (1999). The ethics of representation and action in virtual reality. *Ethics and Information Technology, 1*(1), 5–14.
Brey, P. (2000). Disclosive computer ethics. *ACM Sigcas Computers and Society, 30*(4), 10–16.
Brey, P. A. (2012). Anticipating ethical issues in emerging IT. *Ethics and Information Technology, 14*(4), 305–317.
Broadie, S. (1991). *Ethics with Aristotle*. Oxford University Press.
Bruening, W. H. (1971). Moore and "is-ought". *Ethics, 81*(2), 143–149.
Buxton, M., Phillippi, J. C., & Collins, M. R. (2015). Simulation: A new approach to teaching ethics. *Journal of Midwifery & Women's Health, 60*(1), 70–74.
Carrozzino, M., & Bergamasco, M. (2010). Beyond virtual museums: Experiencing immersive virtual reality in real museums. *Journal of Cultural Heritage, 11*(4), 452–458.
Caspary, W. R. (2006). Dewey and Sartre on ethical decisions: Dramatic rehearsal versus radical choice. *Transactions of the Charles S. Peirce Society, 42*(3), 367–393.
Cha, M., Han, S., Lee, J., & Choi, B. (2012). A virtual reality based fire training simulator integrated with fire dynamics data. *Fire Safety Journal, 50*, 12–24.
Charnley-Parry, I., Whitton, J., Rowe, G., Konrad, W., Meyer, J.-H., Cotton, M. D., Enander, A., Espluga, J., Medina, B., & Bergmans, A. (2017). *Principle for effective engagement. D5.1 for the history of nuclear energy and society project*, Brussels: European Commission.
Cheok, A. D., Tewell, J., Pradana, G. A., & Tsubouchi, K. (n.d.). Touch, taste, and smell: Multi-sensory entertainment. *International Conference on Advances in Computer Entertainment Technology*: Springer, 516–518.
Cherryholmes, C. (1999). *Reading pragmatism*. Teachers College Press.
Chesney, R., & Keats Citron, D. (2018). Deep fakes: A looming challenge for privacy, democracy, and national security. In 107 California Law Review (2019), F.U.o.T.L., Public Law Research Paper No. 692; U of Maryland Legal Studies Research Paper No. 2018–21. (ed.).
Chismar, D. (1988). Empathy and sympathy: The important difference. *The Journal of Value Inquiry, 22*(4), 257–266.
Cialdini, R. B., Brown, S. L., Lewis, B. P., Luce, C., & Neuberg, S. L. (1997). Reinterpreting the empathy-altruism relationship: When one into one equals oneness. *Journal of Personality and Social Psychology, 73*(3), 481.
Climate Assembly UK. (2020). *The path to net zero*. House of Commons with involve, Sortition Foundation and mySociety.
Clohesy, A. M. (2013). *Politics of empathy: Ethics, solidarity, recognition*. Routledge.

Cogburn, J., & Silcox, M. (2014). Against brain-in-a-vatism: On the value of virtual reality. *Philosophy & Technology, 27*(4), 561–579.
Collier, J. (2006). The art of moral imagination: Ethics in the practice of architecture. *Journal of Business Ethics, 66*, 307–317.
Collier, M. (2010). Hume's theory of moral imagination. *History of Philosophy Quarterly, 27*(3), 255–273.
Collingridge, D. (1980). *The social control of technology*. Pinter.
Cotton, M. (2009a). Ethical assessment in radioactive waste management: A proposed reflective equilibrium-based deliberative approach. *Journal of Risk Research, 12*(5), 603–618.
Cotton, M. (2009b). Evaluating the 'ethical matrix' as a radioactive waste management deliberative decision-support tool. *Environmental Values, 18*(2), 153–176.
Cotton, M. (2010). Discourse, upstream public engagement and the governance of human life extension research. *Poiesis & Praxis, 7*(1–2), 135–150.
Cotton, M. (2013). Deliberating intergenerational environmental equity: A pragmatic, future studies approach. *Environmental Values, 22*(3), 317–337.
Cotton, M. (2014a). Ethical matrix and agriculture. In P. B. Thompson & D. M. Kaplan (Eds.), *Encyclopedia of food and agricultural ethics* (pp. 1–10). Springer Netherlands.
Cotton, M. (2014b). *Ethics and technology assessment: A participatory approach*. Springer-Verlag.
Cotton, M. (2017). *Nuclear waste politics: An incrementalist perspective*. Routledge.
Crone, R. A. (1992). The history of stereoscopy. *Documenta Ophthalmologica, 81*(1), 1–16.
Curzer, H. J. (2012). *Aristotle and the virtues*. Oxford University Press.
d'Arms, J., & Jacobson, D. (2000). The moralistic fallacy: On the 'appropriateness' of emotions. *Philosophy and Phenomenological Research, 61*(1), 65–90.
Daft, R. L., & Lengel, R. H. (1986). Organizational information requirements, media richness and structural design. *Management Science, 32*(5), 554–571.
Dalcourt, G. J. (1983). *The methods of ethics*. University Press of America.
Daniels, N. (1979). Wide reflective equilibrium and theory acceptance in ethics. *Journal of Philosophy, 76*(5), 256–282.
Daniels, N. (1996). *Justice and justification: Reflective equilibrium in theory and practice*. Cambridge University Press.
De la Peña, N., Weil, P., Llobera, J., Giannopoulos, E., Pomés, A., Spanlang, B., Friedman, D., Sanchez-Vives, M. V., & Slater, M. (2010). Immersive journalism: Immersive virtual reality for the first-person experience of news. *Presence: Teleoperators and Virtual Environments, 19*(4), 291–301.
De Oliveira, E. C., Bertrand, P., Lesur, M. E. R., Palomo, P., Demarzo, M., Cebolla, A., Baños, R., & Tori, R. (n.d.). Virtual body swap: A new feasible

tool to be explored in health and education. *2016 XVIII Symposium on Virtual and Augmented Reality (SVR)*: IEEE, 81–89.

De Vignemont, F., & Singer, T. (2006). The empathic brain: How, when and why? *Trends in Cognitive Sciences, 10*(10), 435–441.

Deblonde, M., De Graafe, R., & Brom, F. (2007). An ethical toolkit for food companies: Reflections on its use. *Journal of Agricultural and Environmental Ethics, 20*, 99–118.

Decety, J., & Jackson, P. L. (2004). The functional architecture of human empathy. *Behavioral and Cognitive Neuroscience Reviews, 3*(2), 71–100.

Delgado, A., Kjølberg, K. L., & Wickson, F. (2011). Public engagement coming of age: From theory to practice in sts encounters with nanotechnology. *Public Understanding of Science, 20*(6), 826–845.

DeLisi, M., Vaughn, M. G., Gentile, D. A., Anderson, C. A., & Shook, J. J. (2013). Violent video games, delinquency, and youth violence: New evidence. *Youth Violence and Juvenile Justice, 11*(2), 132–142.

Devon, R. (2004). Towards a social ethics of technology: A research prospect. *Techné: Research in Philosophy and Technology, 8*(1), 99–115.

Dewey, J. (1917). *Creative intelligence*. Henry Holt and Co.

Dewey, J. (1982). The pattern of inquiry. In H. S. Thayer (Ed.), *Pragmatism: The classic writings*. Hackett.

Dewey, J., & Tufys, J. H. (1923). *Ethics: Second edition*. Library of Alexandria.

Dinh, H. Q., Walker, N., Hodges, L. F., Song, C., & Kobayashi, A. (n.d.). Evaluating the importance of multi-sensory input on memory and the sense of presence in virtual environments. *Proceedings IEEE Virtual Reality (Cat. No. 99CB36316)*: IEEE, 222–228.

Doppelt, G. (2002). Can traditional ethical theory meet the challenges of feminism, multicuturalism, and environmentalism? *Journal of Ethics, 6*, 383–405.

Dorst, K., & Royakkers, L. (2006). The design analogy: A model for moral problem solving. *Design Studies, 27*(6), 633–656.

Dunn, R. (2004). Moral psychology and expressivism. *European Journal of Philosophy, 12*(2), 178–198.

Edel, A. (1998). *Science and the structure of ethics*. Transaction Publishers.

Ehrsson, H. H., Holmes, N. P., & Passingham, R. E. (2005). Touching a rubber hand: Feeling of body ownership is associated with activity in multisensory brain areas. *Journal of Neuroscience, 25*(45), 10564–10573.

Eisenberg, N., & Fabes, R. A. (1990). Empathy: Conceptualization, measurement, and relation to prosocial behavior. *Motivation and Emotion, 14*(2), 131–149.

Emmelkamp, P. M., Krijn, M., Hulsbosch, A., De Vries, S., Schuemie, M. J., & van der Mast, C. A. (2002). Virtual reality treatment versus exposure in vivo: A comparative evaluation in acrophobia. *Behaviour Research and Therapy, 40*(5), 509–516.

ERC. (2004). *PLUS—A process for ethical decision making*. Washington: Ethics Resource Centre. Retrieved December 1, 2004, from http://www.ethics.org/plus_model.htm

EthicsGame. (2018). *EthicsGame: How it works*. EthicsGame. https://www.ethicsgame.com/exec/site/How_it_works.html

European Commission. (2018). *Responsible research and innovation*. https://ec.europa.eu/programmes/horizon2020/en/h2020-section/responsible-research-innovation

Farrow, R., & Iacovides, I. (2014). Gaming and the limits of digital embodiment. *Philosophy & Technology, 27*(2), 221–233.

Felt, U., & Fochler, M. (2008). The bottom-up meanings of the concept of public participation in science and technology. *Science and Public Policy, 35*(7), 489–499.

Ferreira, M. J. (1994). Hume and imagination: Sympathy and 'the other'. *International Philosophical Quarterly, 34*(1), 39–57.

Fesmire, S. (1994). Educating the moral artist: Dramatic rehearsal in moral education. *Studies in Philosophy and Education, 13*(3–4), 213–227.

Fesmire, S. (2003). *John Dewey and moral imagination: Pragmatism in ethics*. Indiana University Press.

Fesmire, S. (n.d.). Imagination in pragmatist ethics. *Society for the Advancement of American Philosophy 28th Annual Meeting*, University of Nevada, Las Vegas.

Festenstein, M. (1997). *Pragmatism and political theory*. Polity.

Fiorino, D. (1990). Citizen participation and environmental risk: A survey of institutional mechanisms. *Science, Technology & Human Values, 15*(2), 226–243.

Fisher, S. S., Wenzel, E. M., Coler, C., & McGreevy, M. W. (n.d.). Virtual interface environment workstations. *Proceedings of the Human Factors Society Annual Meeting*: SAGE Publications, 91–95.

Food Ethics Council. (2005). *Ethical matrix: Uses*. Food Ethics Council. Retrieved August 2, 2007, from http://www.foodethicscouncil.org/ourwork/tools/ethicalmatrix/uses

Ford, P. J. (2001). A further analysis of the ethics of representation in virtual reality: Multi-user environments. *Ethics and Information Technology, 3*(2), 113–121.

Forester, J. (1984). Bounded rationality and the politics of muddling through. *Public Administration Review, 44*(1), 23–31.

Forester-Miler, H., & Davis, T. (1996). *A practitioner's guide to ethical decision making*. American Counselling Association.

Forsberg, E. M. (2007). Pluralism, the ethical matrix, and coming to conclusions. *Journal of Agricultural and Environmental Ethics, 20*(4), 455–468.

Förster, J., Friedman, R. S., & Liberman, N. (2004). Temporal construal effects on abstract and concrete thinking: Consequences for insight and creative cognition. *Journal of Personality and Social Psychology, 87*(2), 177.

Franklin Institute. (2019). *History of virtual reality*. Philadelphia: The Franklin Institute. Retrieved July 20, 2019, from https://www.fi.edu/virtual-reality/history-of-virtual-reality

Freeman, D., Reeve, S., Robinson, A., Ehlers, A., Clark, D., Spanlang, B., & Slater, M. (2017). Virtual reality in the assessment, understanding, and treatment of mental health disorders. *Psychological Medicine, 47*(14), 2393–2400.

Freina, L., & Ott, M. (2015). A literature review on immersive virtual reality in education: State of the art and perspectives. *eLearning & Software for Education, 1*.

Frey, R. G., & Wellman, C. H. (2008). *A companion to applied ethics*. John Wiley & Sons.

Gaffney, P. (2010). *The force of the virtual: Deleuze, science, and philosophy*. University of Minnesota Press.

Garcia-Betances, R. I., Jiménez-Mixco, V., Arredondo, M. T., & Cabrera-Umpiérrez, M. F. (2015). Using virtual reality for cognitive training of the elderly. *American Journal of Alzheimer's Disease & Other Dementias, 30*(1), 49–54.

Garner, R. T., & Rosen, B. (1967). *Moral philosophy: A systematic introduction to normative ethics and meta-ethics*. Macmillan.

Gaus, G. F. (2002). What is deontology? Part One: Orthodox views. *Journal of Value Inquiry, 35*, 27–42.

Gentile, D. A., Anderson, C. A., Yukawa, S., Ihori, N., Saleem, M., Ming, L. K., Shibuya, A., Liau, A. K., Khoo, A., & Bushman, B. J. (2009). The effects of prosocial video games on prosocial behaviors: International evidence from correlational, longitudinal, and experimental studies. *Personality and Social Psychology Bulletin, 35*(6), 752–763.

Gilligan, C. (1982). *In a different voice: Psychological theory and women's development*. Harvard University Press.

Glantz, K., Durlach, N. I., Barnett, R. C., & Aviles, W. A. (1996). Virtual reality (VR) for psychotherapy: From the physical to the social environment. *Psychotherapy: Theory, Research, Practice, Training, 33*(3), 464.

Gold, J. I., Kant, A. J., & Kim, S. H. (n.d.). Virtual anesthesia: The use of virtual reality for pain distraction during acute medical interventions. *Seminars in Anesthesia, Perioperative Medicine and Pain*: Elsevier, 203–210.

Greenwald, A. G., & Krieger, L. H. (2006). Implicit bias: Scientific foundations. *California Law Review, 94*(4), 945–967.

Greitemeyer, T., & Osswald, S. (2010). Effects of prosocial video games on prosocial behavior. *Journal of Personality and Social Psychology, 98*(2), 211–221.

Grey, T. (1998). Freestanding legal pragmatism. In M. Dickstein (Ed.), *The Revival of pragmatism: New essays on social thought, law, and culture* (pp. 254–274). Duke University Press.

Hacking, I. (Ed.). (1981). *Scientific revolutions*. Oxford University Press.
Haidt, J. (2001). The emotional dog and its rational tail: A social intuitionist approach to moral judgment. *Psychological Review, 108*, 814–834.
Haidt, J. (2003). The moral emotions. In R. J. Davidson, K. R. Scherer, & H. H. Goldsmith (Eds.), *Handbook of affective sciences* (pp. 852–870). Oxford University Press.
Haldane, J. (2012). *Practical philosophy: Ethics, society and culture*. Imprint Academic.
Hamington, M. (2010). Care ethics, John Dewey's 'dramatic rehearsal,' and moral education. *Philosophy of Education Archive*, pp. 121–128.
Haney, K. (1994). Empathy and ethics. *Southwest Philosophy Review, 10*(1), 57–65.
Harper, D. (2002). Talking about pictures: A case for photo elicitation. *Visual Studies, 17*(1), 13–26.
Harris, J. (2006). *The value of life: An introduction to medical ethics*. Routledge.
Harris, S. (2011). *The moral landscape: How science can determine human values*. Simon and Schuster.
Harris, D. (2018). Deepfakes: False pornography is here and the law cannot protect you. *Duke Law & Technology Review, 17*(1), 99–127.
Harvey, D. (1999). Time-space compression and the postmodern condition. *Modernity: Critical Concepts, 4*, 98–118.
Hasen, R. L. (2019). *Deep fakes, bots, and siloed justices: American election law in a post-truth world*, Irvine CA: UC Irvine School of Law Research Paper No. 2019–36.
Healy, T. (2012). The unanticipated consequences of technology. *Nanotechnology: Ethical and social Implications*, pp. 155–173.
Heath, E. (1995). The commerce of sympathy: Adam Smith on the emergence of morals. *Journal of the History of Philosophy, 33*(3), 447–466.
Hedberg, J., & Alexander, S. (1994). Virtual reality in education: Defining researchable issues. *Educational Media International, 31*(4), 214–220.
Heilig, M. L. (1962). *Sensorama simulator*. Google Patents.
Herrero, P., & De Antonio, A. (2005). Intelligent virtual agents keeping watch in the battlefield. *Virtual Reality, 8*(3), 185–193.
Hettinger, L. J., & Riccio, G. E. (1992). Visually induced motion sickness in virtual environments. *Presence: Teleoperators & Virtual Environments, 1*(3), 306–310.
Hickman, L. (2001). *Philosophical tools for technological culture: Putting pragmatism to work*. Indiana University Press.
Hilligoss, B. (2014). Selling patients and other metaphors: A discourse analysis of the interpretive frames that shape emergency department admission handoffs. *Social Science & Medicine, 102*, 119–128.
Hodgkinson, G. (2016). Lock up your stories-here comes virtual reality. *TECHART: Journal of Arts and Imaging Science, 3*(4), 10–14.

Hoffman, M. (2000). *Empathy and moral development.* Cambridge University Press.
Hoffman, M. L. (2008). Empathy and prosocial behaviour. In M. Lewis, J. M. Haviland-Jones, & L. Feldman-Barrett (Eds.), *Handbook of emotions* (pp. 440–455). Guildford Press.
Hoffman, H. G., Patterson, D. R., & Carrougher, G. J. (2000). Use of virtual reality for adjunctive treatment of adult burn pain during physical therapy: A controlled study. *The Clinical Journal of Pain, 16*(3), 244–250.
Hoffman, H. G., Chambers, G. T., Meyer, W. J., III, Arceneaux, L. L., Russell, W. J., Seibel, E. J., Richards, T. L., Sharar, S. R., & Patterson, D. R. (2011). Virtual reality as an adjunctive non-pharmacologic analgesic for acute burn pain during medical procedures. *Annals of Behavioral Medicine, 41*(2), 183–191.
Horlick-Jones, T., Walls, J., Rowe, G., Pidgeon, N., Poortinga, W., & O'Riordan, T. (2006). On evaluating the GM Nation? Public debate about the commercialisation of transgenic crops in Britain. *New Genetics and Society, 25*(3), 265–288.
How virtual reality can create the ultimate empathy machine. (2015). TED Talk, 2015.
Howard, B. J., Forsberg, E. M., Kaiser, M., & Oughton, D. (n.d.). An ethical dimension to sustainable restoration and long-term management of contaminated areas. *International Conference On Radioactivity in The Environment*, Monaco, 506–510.
Hume, D. (1739). *Treatise on human nature: Of virtue and vice in general.* Oxford University Press.
Husserl, E. (1931). *Cartesian meditations.* Martinus Nijhoff.
Hutcheson, F. (1725). *An essay on the nature and conduct of the passions and affections. With illustrations on the moral sense. By the author of the inquiry into the original of our ideas of beauty and virtue.* London: J. and J. Knapton, John Darby, Thomas Osborne, Jauton Gilliver, John Crownfield.
IDC. (2018). *Worldwide quarterly augmented and virtual reality headset tracker.* International Data Corporation. https://www.idc.com/tracker/showproductinfo.jsp?prod_id=1501
Isaak, J., & Hanna, M. J. (2018). User data privacy: Facebook, Cambridge Analytica, and privacy protection. *Computer, 51*(8), 56–59.
James, W. (1907). *Pragmatism: A new name for some old ways of thinking.* Longmans, Green and Co.
James, W. (1976). *Pragmatism.* Harvard University Press.
James, W. (1978). The meaning of truth. In F. H. Buckhardt (Ed.), *Essays in philosophy: The works of William James.* Harvard University Press.
Johnson, M. (1993). *Moral Imagination: Implications of cognitive science for ethics.* University of Chicago Press.
Jonathan, P. J. Y., Fung, C. C., & Wong, K. W. (2009). Devious chatbots-interactive malware with a plot. In J.-H. Kim, S. S. Ge, P. Vadakkepat,

J. Norbert, A. A. Manum, K. Sadasivan Puthusserypady, U. Rückert, J. Sitte, U. Witkowski, R. Nakatsu, T. Braun, J. Baltes, J. R. Anderson, C.-C. Wong, I. Verner, & D. Ahlgren (Eds.), *Progress in robotics. FIRA 2009. Communications in computer and information science, vol 44* (pp. 110–118). Springer.

Jones, T. M. (1991). Ethical decision making by individuals in organizations: An issue-contingent model. *The Academy of Management Review, 16*(2), 366–395.

Jones, S. (2017). Disrupting the narrative: Immersive journalism in virtual reality. *Journal of Media Practice, 18*(2–3), 171–185.

Joss, S., & Brownlea, A. (1999). Considering the concept of procedural justice for public policy-and decision-making in science and technology. *Science and Public Policy, 26*(5), 321–330.

Juul, J. (2010). The game, the player, the world: Looking for a heart of gameness. *PLURAIS-Revista Multidisciplinar, 1*(2), 248–270.

Kahneman, D. (2003). A perspective on judgment and choice: Mapping bounded rationality. *American Psychologist, 58*(9), 697–720.

Kahneman, D., & Tversky, A. (1984). Choices, values, and frames. *American Psychologist, 39*(4), 341.

Kaiser, M., & Forsberg, E. M. (2001). Assessing fisheries—Using an ethical matrix in participatory processes. *Journal of Agricultural and Environmental Ethics, 14*, 191–200.

Kaiser, M., Millar, K., Forsberg, E.-M., Baune, O., Mepham, B., Thorstensen, E., & Tomkins, S. (2004). Decision-making frameworks. In V. Beekman (Ed.), *Evaluation of ethical bio-technology assessment tools for agriculture and food production: Interim report ethical bio-ta tools.* Agricultural Economics Research Institute.

Kaiser, M., Millar, K., Forsberg, E. M., Thorstensen, E., & Tomkins, S. (2007). Developing the ethical matrix as a decision support framework: GM fish as a case study. *Journal of Agricultural and Environmental Ethics, 20*(1), 53–63.

Kaler, J. (1999). What's the good of ethical theory? *Business Ethics: A European Review, 8*(4), 206–213.

Kaplan, A. M., & Haenlein, M. (2009). The fairyland of second life: Virtual social worlds and how to use them. *Business Horizons, 52*(6), 563–572.

Kavathatzopoulos, I. (2003). The use of information and communication technology in the training for ethical competence in business. *Journal of Business Ethics, 48*(1), 43–51.

Kennedy, R. S., Drexler, J., & Kennedy, R. C. (2010). Research in visually induced motion sickness. *Applied Ergonomics, 41*(4), 494–503.

Keulartz, J., Korthals, M., Schermer, M., & Swierstra, T. E. (Eds.). (2002). *Pragmatist ethics for a technological culture.* Kluwer.

Keulartz, J., Shermer, M., Korthals, M., & Swierstra, T. (2004). Ethics in technological culture: A programmatic proposal for a pragmatist approach. *Science, Technology & Human Values, 29*(1), 3–29.

Kilteni, K., Groten, R., & Slater, M. (2012). The sense of embodiment in virtual reality. *Presence: Teleoperators and Virtual Environments, 21*(4), 373–387.

Kinateder, M., Ronchi, E., Nilsson, D., Kobes, M., Müller, M., Pauli, P., & Mühlberger, A. (n.d.). Virtual reality for fire evacuation research. *2014 Federated Conference on Computer Science and Information Systems*: IEEE, 313–321.

Kleinman, D. L. (Ed.). (2000). *Science, technology and democracy*. State University of New York Press.

Klinger, E., Bouchard, S., Légeron, P., Roy, S., Lauer, F., Chemin, I., & Nugues, P. (2005). Virtual reality therapy versus cognitive behavior therapy for social phobia: A preliminary controlled study. *Cyberpsychology & Behavior, 8*(1), 76–88.

Kohlberg, L. (1984). *The psychology of moral development : The nature and validity of moral stages*. Harper & Row.

Konrad, W., Espluga, J., Bergmans, A., Charnley-Parry, I., Cotton, M. D., Enander, A., Meyer, J.-H., Rowe, G., & Whitton, J. (2018). *Comparative cross-country analysis on preliminary identification of key factors underlying public perception and societal engagement with nuclear developments in different national contexts*. Brussels: European CommissionD4.2 (2018 update).

Konrath, S. H., O'Brien, E. H., & Hsing, C. (2011). Changes in dispositional empathy in American college students over time: A meta-analysis. *Personality and Social Psychology Review, 15*(2), 180–198.

Kool, H. (2016). The ethics of immersive journalism: A rhetorical analysis of news storytelling with virtual reality technology. *Intersect: The Stanford Journal of Science, Technology, and Society, 9*(3). http://ojs.stanford.edu/ojs/index.php/intersect/article/view/871/863

Kors, M. J., Ferri, G., Van Der Spek, E. D., Ketel, C., & Schouten, B. A. (n.d.). A breathtaking journey. On the design of an empathy-arousing mixed-reality game. *Proceedings of the 2016 Annual Symposium on Computer-Human Interaction in Play*: ACM, 91–104.

Krabbenborg, L. (2013). Dramatic rehearsal on the societal embedding of the lithium chip. In S. van der Burg & T. Swierstra (Eds.), *Ethics on the laboratory floor* (pp. 168–187). Springer.

Krueger, M. W. (1993). An easy entry artificial reality. In A. Wexelblat (Ed.), *Virtual reality: Applications and explorations* (pp. 147–161). Elsevier.

Krueger, M. W., Gionfriddo, T., & Hinrichsen, K. (n.d.). VIDEOPLACE—An artificial reality. *ACM SIGCHI Bulletin*: ACM, 35–40.

Kuhn, J. W. (1998). Emotion as well as reason: Getting students beyond 'interpersonal accountability'. *Journal of Business Ethics, 17*(3), 295–308.

LaFollette, H. (2003). Pragmatic ethics. In H. LaFollette (Ed.), *Ethical theory*. Blackwell Publishing.

Laterza, V. (2018). Cambridge Analytica, independent research and the national interest. *Anthropology Today, 34*(3), 1–2.

Law, A., & Mooney, G. (2012). Competitive nationalism: State, class, and the forms of capital in devolved Scotland. *Environment and Planning C: Government and Policy, 30*(1), 62–77.

Lekan, T. (2006). Pragmatist metaethics: Moral theory as deliberative practice. *Southern Journal of Philosophy, 44*(2), 253–272.

Lele, A. (2013). Virtual reality and its military utility. *Journal of Ambient Intelligence and Humanized Computing, 4*(1), 17–26.

Lepouras, G., & Vassilakis, C. (2004). Virtual museums for all: Employing game technology for edutainment. *Virtual Reality, 8*(2), 96–106.

Lévy, P. (1998). *Becoming virtual: Reality in the digital age*. Plenum Press.

Lidskog, R. (2008). Scientised citizens and democratised science. Re-assessing the expert-lay divide. *Journal of Risk Research, 11*(1–2), 69–86.

Lindley, S. E., Le Couteur, J., & Berthouze, N. L. (n.d.). Stirring up experience through movement in game play: Effects on engagement and social behaviour. *Proceedings of the SIGCHI Conference on Human Factors in Computing Systems*: ACM, 511–514.

Lipps, T. (1903). *Aesthetik*. Voss Verlag.

Litzky, B. E. (2012). Review of EthicsGame simulation. *Journal of Business Ethics Education, 9*, 485–488.

Lloyd, P., & van de Poel, I. (2008). Designing games to teach ethics. *Science and Engineering Ethics, 14*(3), 433–447.

London, A. J. (2001). The independence of practical ethics. *Theoretical Medicine and Bioethics, 22*(2), 87–105.

Luppicini, R. (Ed.). (2010). *Technoethics and the evolving knowledge society: Ethical issues in technological design, research, development, and innovation*. IGI Global.

Machlaren, P., & Seedhouse, D. (2001). *Computer mediated communication with integrated graphical tools used for health care decision-making*, October 2004.

MacIntyre, A. (1984a). *After virtue: A study in moral theory*. University of Notre Dame Press.

MacIntyre, A. (1984b). Does applied ethics rest on a mistake? *The Monist, 67*, 489–513.

Mackay, T., Ewing, M., Newton, F., & Windisch, L. (2009). The effect of product placement in computer games on brand attitude and recall. *International Journal of Advertising, 28*(3), 423–438.

Madary, M., & Metzinger, T. K. (2016). Real virtuality: A code of ethical conduct. recommendations for good scientific practice and the consumers of VR-technology. *Frontiers in Robotics and AI, 3*.

Magda, R. M. R. (2001). Transmodernity, neotribalism and postpolitics. *Interlitteraria, 6*(6), 9–18.

Mann, S. (2015). Phenomenal augmented reality: Advancing technology for the future of humanity. *IEEE Consumer Electronics Magazine, 4*(4), 92–97.

Maranta, A., Guggenheim, M., Gisler, P., & Pohl, C. (2003). The reality of experts and the imagined lay person. *Acta Sociologica, 46*(2), 150–165.
Marshall, J. (1999). *An ethical decision-making model: Five steps of principled reasoning*. Josephson Institute of Ethics. http://www.ethicsscoreboard.com/rb_5step.html
Massey, D. (1992). Politics and space/time. *New Left Review, 196*, 65–84.
McCracken, J., & Shaw, B. (1995). Virtue ethics and contractarianism: Towards a reconciliation. *Business Ethics Quarterly*, pp. 297–312.
McGee, M. T. (2001). *Beyond Ballyhoo: Motion picture promotion and Gimmicks.* McFarland.
McNally, H., Howley, P., & Cotton, M. (2018). Public perceptions of shale gas in the UK: Framing effects and decision heuristics. *Energy, Ecology and Environment, 3*(6), 305–316.
McVea, J. F. (2007). Constructing good decisions in ethically charged situations: The role of dramatic rehearsal. *Journal of Business Ethics, 70*, 375–390.
McWilliams, V., & Nahavandi, A. (2006). Using live cases to teach ethics. *Journal of Business Ethics, 67*(4), 421–433.
Mepham, B. (1999a). A framework for the ethical analysis of novel foods: The ethical matrix. *Journal of Agricultural and Environmental Ethics, 12*, 165–176.
Mepham, B. (1999b). A framework for the ethical analysis of novel foods: The ethical matrix. *Journal of Agricultural and Environmental Ethics, 12*(2), 165–176.
Michael, M. A. (2003). What's in a name? Pragmatism, essentialism, and environmental ethics. *Environmental Values, 12*, 361–379.
Michael, D., & Chen, S. (2005). *Serious games: Games that educate, train, and inform*. Muska & Lipman.
Milgram, P., Takemura, H., Utsumi, A., & Kishino, F. (1995). Augmented reality: A class of displays on the reality-virtuality continuum. *Telemanipulator and Telepresence Technologies, 2351*, 282–292.
Millar, K., Thorstensen, E., Tomkins, S., Mepham, B., & Kaiser, M. (2007). Developing the ethical delphi. *Journal of Agricultural and Environmental Ethics, 20*, 53–63.
Misak, C. (2000). *Truth, politics, morality.* Routledge.
Moberg, D., & Seabright, M. (2000). The development of moral imagination. *Business Ethics Quarterly, 10*, 845–884.
Mohr, A. (2007). Against the stream: Moving public engagement on nanotechnologies upstream. In R. Flynn & P. Bellamy (Eds.), *Risk and the acceptance of new technologies*. Palgrave Macmillan.
Moore, G. E. (1903). *Principia Ethica*. Cambridge University Press.
Moore, E. C. (1957). The moralistic fallacy. *The Journal of Philosophy, 54*(2), 29–42.
Moula, P., & Sandin, P. (2015). Evaluating ethical tools. *Metaphilosophy, 46*(2), 263–279.

Moulard-Leonard, V. (2008). *Bergson-Deleuze encounters: Transcendental experience and the thought of the virtual*. SUNY Press.

Mujber, T. S., Szecsi, T., & Hashmi, M. S. (2004). Virtual reality applications in manufacturing process simulation. *Journal of Materials Processing Technology, 155*, 1834–1838.

Murphy, J., Levidow, L., & Carr, S. (2006). Regulatory standards for environmental risks: Understanding the US–European Union conflict over genetically modified crops. *Social Studies of Science, 36*(1), 133–160.

Nash, K. (2018). Virtual reality witness: Exploring the ethics of mediated presence. *Studies in Documentary Film, 12*(2), 119–131.

Neff, G., & Nagy, P. (2016). Automation, algorithms, and politics| talking to Bots: Symbiotic agency and the case of Tay. *International Journal of Communication, 10*, 4915–4931.

Nemet, G. F. (2009). Demand-pull, technology-push, and government-led incentives for non-incremental technical change. *Research policy, 38*(5), 700–709.

Nowotny, H. (2003). Democratising expertise and socially robust knowledge. *Science and Public Policy, 30*(3), 151–156.

Nussbaum, M. (1986). *The fragility of goodness: Luck and ethics in Greek tragedy and philosophy*. Cambridge University Press.

Nussbaum, M. (2001). *Upheavals of thought: The intelligence of the emotions*. Cambridge University Press.

O'Brolcháin, F., Jacquemard, T., Monaghan, D., O'Connor, N., Novitzky, P., & Gordijn, B. (2016). The convergence of virtual reality and social networks: Threats to privacy and autonomy. *Science and Engineering Ethics, 22*(1), 1–29.

Olofsson, J. K., Niedenthal, S., Ehrndal, M., Zakrzewska, M., Wartel, A., & Larsson, M. (2017). Beyond smell-o-vision: Possibilities for smell-based digital media. *Simulation & Gaming, 48*(4), 455–479.

Orr, T. J., Mallet, L., & Margolis, K. A. (2009). Enhanced fire escape training for mine workers using virtual reality simulation. *Mining Engineering, 61*(11), 41.

Owen, R., Macnaghten, P., & Stilgoe, J. (2012). Responsible research and innovation: From science in society to science for society, with society. *Science and Public Policy, 39*(6), 751–760.

Page, A. S., Cooper, A. R., Griew, P., & Jago, R. (2010). Children's screen viewing is related to psychological difficulties irrespective of physical activity. *Pediatrics, 126*(5), e1011–e1017.

Palmer, M. T. (1995). Interpersonal communication and virtual reality: Mediating interpersonal relationships. In F. Biocca & M. R. Levy (Eds.), *Communication in the age of virtual reality* (pp. 277–299). Laurence Erlbaum and Associates Ltd.

Park, E.-J. (2012). An integrated ethical decision-making model for nurses. *Nursing Ethics, 19*(1), 139–159.

Parker, K. A. (1996). Pragmatism and environmental thought. In A. Light & E. Katz (Eds.), *Environmental pragmatism*. Routledge.

Paul, S. (2009). Binaural recording technology: A historical review and possible future developments. *Acta Acustica United with Acustica, 95*(5), 767–788.
Peirce, C. S. (1982). Definition and description of pragmatism. In H. S. Thayer (Ed.), *Pragmatism: The classic writings*. Hackett.
Peirce, C. S., & Dewey, J. (2017). How to make our ideas clear. In C. Pierce (Ed.), *Chance, love, and logic* (pp. 32–60). Routledge.
Pellizzoni, L. (2012). Strong will in a messy world. Ethics and the government of technoscience. *NanoEthics, 6*(3), 257–272.
Persily, N. (2017). The 2016 US Election: Can democracy survive the internet? *Journal of Democracy, 28*(2), 63–76.
Pidgeon, N., & Rogers-Hayden, T. (2007). Opening up nanotechnology dialogue with the publics: Risk communication or 'upstream engagement'? *Health, Risk & Society, 9*(2), 191–210.
Porter, E. (1999). *Feminist perspectives on ethics*. Longman.
Portman, M. E., Natapov, A., & Fisher-Gewirtzman, D. (2015). To go where no man has gone before: Virtual reality in architecture, landscape architecture and environmental planning. *Computers, Environment and Urban Systems, 54*, 376–384.
Potter, R. B. (n.d.). The origins and applications of 'potter boxes'. *State of the World Forum*, San Francisco, CA.
Powers, M. B., & Emmelkamp, P. M. (2008). Virtual reality exposure therapy for anxiety disorders: A meta-analysis. *Journal of Anxiety Disorders, 22*(3), 561–569.
Prensky, M. (2006). *Don't bother me, Mom, I'm learning!: How computer and video games are preparing your kids for 21st century success and how you can help!*: Paragon House St. Paul.
Pujol, L. (2004). Archaeology, museums and virtual reality. *Digithum, 6*, 1–9.
Putnam, H. (1994). *Words and life*. Harvard University Press.
Radder, H. (2004). Pragmatism, ethics, and technology'. *Techné: Science, Technology & Human Values, 7*(3), 10–18.
Rauterberg, M. (2004). *Positive effects of entertainment technology on human behaviour. Building the information society*. Springer, pp. 51–58.
Rawls, J. (1951). Outline of a decision procedure for ethics. *The Philosophical Review, 60*(2), 177–197.
Rawls, J. (1999). *A theory of justice* (2nd ed.). Oxford University Press.
Reed, S., Kreylos, O., Hsi, S., Kellogg, L., Schladow, G., Yikilmaz, M., Segale, H., Silverman, J., Yalowitz, S., & Sato, E. (n.d.). Shaping watersheds exhibit: An interactive, augmented reality sandbox for advancing earth science education. *AGU Fall Meeting Abstracts*, abstract id. ED34A-01.
Reger, G. M., Holloway, K. M., Candy, C., Rothbaum, B. O., Difede, J., Rizzo, A. A., & Gahm, G. A. (2011). Effectiveness of virtual reality exposure therapy for active duty soldiers in a military mental health clinic. *Journal of Traumatic Stress, 24*(1), 93–96.

Reznick, R. K., & MacRae, H. (2006). Teaching surgical skills—Changes in the wind. *New England Journal of Medicine, 355*(25), 2664–2669.

Rip, A., & Kemp, R. (1998). Technological change. In S. Rayner & E. Malone (Eds.), *Human choices and climate change*. Battelle.

Rip, A., Schot, J. W., & Misa, T. J. (1995). Constructive technology assessment: A new paradigm for managing technology in society. In A. Rip, J. W. Schot, & T. J. Misa (Eds.), *Managing technology in society. The approach of constructive technology assessment* (pp. 1–12). Pinter Publishers.

Rizzo, A., Pair, J., McNerney, P. J., Eastlund, E., Manson, B., Gratch, J., Hill, R., & Swartout, B. (2005). Development of a VR therapy application for Iraq war military personnel with PTSD. *Studies in Health Technology and Informatics, 111*, 407–413.

Rizzo, A. S., Lange, B., Suma, E. A., & Bolas, M. (2011). Virtual reality and interactive digital game technology: New tools to address obesity and diabetes. *Journal of Diabetes Science and Technology, 5*(2), 256–264.

Rizzo, A., Buckwalter, J. G., John, B., Newman, B., Parsons, T., Kenny, P., & Williams, J. (2012). STRIVE: Stress resilience in virtual environments: A pre-deployment VR system for training emotional coping skills and assessing chronic and acute stress responses. *Studies in Health Technology and Informatics, 173*, 379–385.

Robins, K., & Webster, F. (2003). *Times of the technoculture: From the information society to the virtual life*. Routledge.

Rorty, R. (1995). Is truth a goal of inquiry? Davidson vs. Wright. *Philosophical Quarterly, 45*(189), 281–300.

Rothbaum, B. O., Hodges, L., Smith, S., Lee, J. H., & Price, L. (2000). A controlled study of virtual reality exposure therapy for the fear of flying. *Journal of Consulting and Clinical Psychology, 68*(6), 1020–1026.

Rueda, J., & Lara, F. (2020). Virtual reality and empathy enhancement: Ethical aspects. *Frontiers in Robotics and AI*, 7. https://doi.org/10.3389/frobt.2020.506984

Ryan, M.-L. (2001). *Narrative as virtual reality. Immersion and interactivity in literature*. Johns Hopkins University Press.

Sanchez-Vives, M. V., & Slater, M. (2005). From presence to consciousness through virtual reality. *Nature Reviews Neuroscience, 6*, 332–339.

Satterfield, T. (2001). In search of value literacy: Suggestions for the elicitation of environmental values. *Environmental Values, 10*(3), 331–359.

Schechtman, M. (2012). The story of my (second) life: Virtual worlds and narrative identity. *Philosophy & Technology, 25*(3), 329–343.

Schmidt-Felzmann, H. (2003). Pragmatic principles—Methodological pragmatism in the principle-based approach to bioethics. *Journal of Medicine and Philosophy, 28*(5–6), 581–596.

Schroeder, D., & Palmer, C. (2003). Technology assessment and the 'ethical matrix'. *Poiesis & Praxis, 1*(4), 295–307.

Schroeder, R., Heather, N., & Lee, R. M. (1998). The sacred and the virtual: Religion in multi-user virtual reality. *Journal of Computer-Mediated Communication, 4*(2).

Schultheis, M. T., & Rizzo, A. A. (2001). The application of virtual reality technology in rehabilitation. *Rehabilitation Psychology, 46*(3), 296.

Schumann, K., Zaki, J., & Dweck, C. S. (2014). Addressing the empathy deficit: Beliefs about the malleability of empathy predict effortful responses when empathy is challenging. *Journal of Personality and Social Psychology, 107*(3), 475–493.

Schwienhorst, K. (2002). Why virtual, why environments? Implementing virtual reality concepts in computer-assisted language learning. *Simulation & Gaming, 33*(2), 196–209.

Sclove, R. (1995). *Democracy and technology*. Guilford Publications.

Searle, J. R. (1964). How to derive 'ought' from 'is'. *The Philosophical Review, 73*, 43–48.

Seedhouse, D. (1998). *Ethics: The heart of health care*. Wiley.

Seidel, R. J., & Chatelier, P. R. (2013). *Virtual reality, training's future?: Perspectives on virtual reality and related emerging technologies*. Springer Science & Business Media.

Seymour, N. E., Gallagher, A. G., Roman, S. A., O'brien, M. K., Bansal, V. K., Andersen, D. K., & Satava, R. M. (2002). Virtual reality training improves operating room performance: Results of a randomized, double-blinded study. *Annals of Surgery, 236*(4), 458.

Sharma, S., Lomash, H., & Bawa, S. (2015). Who regulates ethics in the virtual world? *Science and Engineering Ethics, 21*(1), 19–28.

Sheetz, T., Vidal, J., Pearson, T. D., & Lozano, K. (2005). Nanotechnology: Awareness and societal concerns. *Technology in Society, 27*(3), 329–345.

Sherman, B., & Judkins, P. (1992). *Glimpses of heaven, visions of hell: Virtual reality and its implications*. Hodder & Stoughton.

Shin, D.-H. (2017). The role of affordance in the experience of virtual reality learning: Technological and affective affordances in virtual reality. *Telematics and Informatics, 34*(8), 1826–1836.

Shrage, L. (1994). Interpretative ethics, cultural relativism and feminist theory. In L. Shrage (Ed.), *Moral dilemmas of feminism* (pp. 162–184). Routledge.

Singer, P. A. (2000). Medical ethics. *British Medical Journal, 321*(7256), 282–285.

Sinnott-Armstrong, W. (1987). Moral realisms and moral dilemmas. *The Journal of Philosophy, 84*(263–276).

Sirkkunen, E., & Uskali, T. (2019). Virtual reality journalism. In *The international encyclopedia of journalism studies*. Wiley Blackwell.

Slater, M., & Usoh, M. (1993). Representations systems, perceptual position, and presence in immersive virtual environments. *Presence: Teleoperators & Virtual Environments, 2*(3), 221–233.
Slote, M. (2007). *The ethics of care and empathy*. Routledge.
Slote, M. (2017). The many faces of empathy. *Philosophia, 45*(3), 843–855.
Smith, L. G. (1987). The evolution of public participation in Canada: Implications for participatory practice. *British Journal of Canadian Studies, 2*(2), 213–235.
Songhorian, S. (2019). The contribution of empathy to ethics. *International Journal of Philosophical Studies, 27*(2), 244–264.
Spencer, A. R. (2013). The dialogues as dramatic rehearsal: Plato's republic and the moral accounting metaphor. *The Pluralist, 8*(2), 26–35.
Spiegel, J. S. (2018). The ethics of virtual reality technology: Social hazards and public policy recommendations. *Science and Engineering Ethics, 24*(5), 1537–1550.
Spielthenner, G. (2005). Consequentialism or deontology? *Philosophia, 33*(1), 217–235.
Stein, E. (1917). *On the problem of empathy*. ICS Publishers.
Steuer, J. (1994). Defining virtual reality: Dimensions determining telepresence. *Journal of Communication, 42*(4), 73–93.
Stirling, A., O'Donovan, C., & Ayre, B. (2018). Which way? Who says? Why? Questions on the multiple directions of social progress. *Technology's Stories*, pp. 1–20.
Strikwerda, L. (2015). Present and future instances of virtual rape in light of three categories of legal philosophical theories on rape. *Philosophy & Technology, 28*(4), 491–510.
Sturgis, P., & Allum, N. (2004). Science in society: Re-evaluating the deficit model of public attitudes. *Public Understanding of Science, 13*(1), 55–74.
Suler, J. (2004). The online disinhibition effect. *Cyberpsychology & Behavior, 7*(3), 321–326.
Sumner, L. W. (1967). Normative ethics and metaethics. *Ethics, 77*(2), 95–106.
Sutherland, I. E. (1965/2002). The ultimate display. In R. Packer & K. Jordan (Eds.), *Multimedia: From Wagner to virtual reality*. WW Norton & Company.
Taebi, B., Correlje, A., Cuppen, E., Dignum, M., & Pesch, U. (2014). Responsible innovation as an endorsement of public values: The need for interdisciplinary research. *Journal of Responsible Innovation, 1*(1), 118–124.
Terry, C., & Cain, J. (2016). The emerging issue of digital empathy. *American Journal of Pharmaceutical Education, 80*(4), 58.
The World Economic Forum. (2020). *Our mission*. Geneva. Retrieved June 12, 2020, from https://www.weforum.org/about/world-economic-forum/
Thompson, D. F. (2007). *What is practical ethics?* Harvard University.
Thomson, A. (1999). *Critical reasoning in ethics: A practical introduction*. Routledge.

Torick, E. (1998). Highlights in the history of multichannel sound. *Journal of the Audio Engineering Society, 46*(1/2), 27–31.

Trope, Y., & Liberman, N. (2010). Construal-level theory of psychological distance. *Psychological Review, 117*(2), 440.

UNVR. (2020). *UN virtual reality.* New York: United Nations Virtual Reality (UNVR), A Project Implemented by the UN SDG Action Campaign. Retrieved June 12, 2020, from http://unvr.sdgactioncampaign.org/vr-films/#.YATy5y-l2cY

Urquhart, C., Underhill-Sem, Y., Pace, T., Houssian, A., & McArthur, V. (2009). Are socially exclusive values embedded in the avatar creation interfaces of MMORPGs? *Journal of Information, Communication and Ethics in Society, 7*(2/3), 192–210.

Vallor, S. (2016). *Technology and the virtues: A philosophical guide to a future worth wanting.* Oxford University Press.

van der Scheer, L., & Widdershoven, G. (2004). Integrated empirical ethics: Loss of normativity? *Medicine, Health Care and Philosophy, 7*(1), 71–79.

Van De Poel, I. (2001). Ethics, technology assessment and industry. *TA-Datenbank-Nachrichten, 2*(10), 51–61.

Van Den Ende, J., Mulder, K., Knot, M., Moors, E., & Vergragt, P. (1998). Traditional and modern technology assessment: Toward a toolkit. *Technological Forecasting and Social Change, 58*(1), 5–21.

Van Wynsberghe, A., & Robbins, S. (2014). Ethicist as Designer: A pragmatic approach to ethics in the lab. *Science and Engineering Ethics, 20*(4), 947–961.

Van-Hoose, W. H. (1980). Ethics and counseling. *Counseling & Human Development, 13*(1), 1–12.

Verbeek, P.-P. (2006). Materializing morality: Design ethics and technological mediation. *Science, Technology, & Human Values, 31*(3), 361–380.

Verbeek, P.-P. (2011). *Moralizing technology: Understanding and designing the morality of things.* University of Chicago Press.

Vitell, S. J., & Nin-Ho, F. (1997). Ethical decision making in marketing: A synthesis and evaluation of scales measuring the various components of decision making in ethical situations. *Journal of Business Ethics, 16*(7), 699–717.

Wade, N., & Ono, H. (1985). The stereoscopic views of Wheatstone and Brewster. *Psychological Research, 47*(3), 125–133.

Walker, M. U. (1989). Moral understandings: A feminist study in ethics. *Hypatia, 4*(2), 15–28.

Wankel, C., & Malleck, S. (2010). Exploring new ethical issues in the virtual worlds of the twenty-first century. In C. Wankel & S. Malleck (Eds.), *Emerging ethical issues of life in virtual worlds* (pp. 1–14). Information Age Publishing.

Wann, J., & Mon-Williams, M. (1996). What does virtual reality NEED?: Human factors issues in the design of three-dimensional computer environments. *International Journal of Human-Computer Studies, 44*(6), 829–847.

Ward, K. (2018). Social networks, the 2016 US presidential election, and Kantian ethics: Applying the categorical imperative to Cambridge Analytica's behavioral microtargeting. *Journal of Media Ethics, 33*(3), 133–148.
Wehling, P. (2012). From invited to uninvited participation (and back?): Rethinking civil society engagement in technology assessment and development. *Poiesis & Praxis, 9*(1–2), 43–60.
Wenk, E. (1975). Technology assessment in public policy: A new instrument for social management of technology. *Proceedings of the IEEE, 63*(3), 371–379.
Werhane, P. H. (2015). Moral imagination. *Wiley encyclopedia of management*, 2(1–2), John Wiley & Sons, Ltd. https://doi.org/10.1002/9781118785317.weom020036
Weston, A. (2000). *A 21st century ethical toolbox*. Oxford University Press.
Wethington, H., Pan, L., & Sherry, B. (2013). The association of screen time, television in the bedroom, and obesity among school-aged youth: 2007 National Survey of Children's Health. *Journal of School Health, 83*(8), 573–581.
Whitton, J., Brasier, K., Parry, I., & Cotton, M. (2017). The development of shale gas governance in the United Kingdom and United States: Opportunities for public participation and implications for social justice. *Energy Research & Social Science, 26*, 11–22.
Widdershoven, G., & Van der Scheer, L. (2008). Theory and methodology of empirical ethics: A pragmatic hermeneutic perspective. In G. Widdershoven, T. Hope, J. McMillan, & L. Van der Scheer (Eds.), *Empirical ethics in psychiatry* (pp. 23–36). Oxford University Press.
Wiener, P. P. (1974). Pragmatism. In P. P. Wiener (Ed.), *The dictionary of the history of ideas: Studies of selected pivotal ideas* (pp. 551–570). Charles Scribner's Sons.
Wilbur, S. P. (2013). An archaeology of cyberspaces: Virtuality, community, identity. In D. Porter (Ed.), *Internet culture* (pp. 5–22). Routledge.
Willaert, W. I., Aggarwal, R., Van Herzeele, I., Cheshire, N. J., & Vermassen, F. E. (2012). Recent advancements in medical simulation: Patient-specific virtual reality simulation. *World Journal of Surgery, 36*(7), 1703–1712.
Wilsdon, J., & Willis, R. (2004). *See-through science: Why public engagement needs to move upstream*. Demos.
Wilson, J. R. (1996). Effects of participating in virtual environments a review of current knowledge. *Safety Science, 23*(1), 39–51.
Winkler, E. R., & Coombs, J. R. (1993). *Applied ethics: A reader*. Blackwell.
Wolfendale, J. (2007). My avatar, my self: Virtual harm and attachment. *Ethics and Information Technology, 9*(2), 111–119.
Won, A. S., Bailenson, J., & Lanier, J. (2015). Homuncular flexibility: The human ability to inhabit nonhuman avatars. In R. A. Scott, S. M. Kosslyn, & M. Buchmann (Eds.), *Emerging trends in the social and behavioral science: An interdisciplinary, searchable, and linkable resource* (pp. 1–16). John Wiley & Sons.

Wynne, B. (1996). May the sheep safely graze? A reflexive view of the expert-lay knowledge divide. In S. Lash, B. Szerszynski, & B. Wynne (Eds.), *Risk, environment and modernity*. Sage Publications.

Yee, N., Bailenson, J. N., Urbanek, M., Chang, F., & Merget, D. (2007). The unbearable likeness of being digital: The persistence of nonverbal social norms in online virtual environments. *CyberPsychology & Behavior, 10*(1), 115–121.

Zahn-Waxler, C., Hollenbeck, B., & Radke-Yarrow, M. (1985). The origins of empathy and altruism. In M. W. Fox & L. Mickley (Eds.), *Advances in animal welfare science 1984* (pp. 21–41). Springer.

Index

M. Cotton, *Virtual Reality, Empathy and Ethics*, https://doi.org/10.1007/978-3-030-72907-3

The manufacturer's authorised representative in the EU is Springer Nature Customer Service Centre GmbH, Europaplatz 3, 69115 Heidelberg, Germany. If you have any concerns regarding our products, please contact ProductSafety@springernature.com

Printed and bound by CPI Group (UK) Ltd, Croydon, CR0 4YY
30/09/2026
02225666-0015